Domestic Fowl and Ornamental Poultry

by H.D. Richardson

with an introduction by Jackson Chambers

This work contains material that was originally published in 1856.

This publication is within the Public Domain.

This edition is reprinted for educational purposes and in accordance with all applicable Federal Laws.

Introduction Copyright 2017 by Jackson Chambers

COVER CREDITS

Front Cover
From Interior Image

Back Cover
From Interior Image

Research / Sources
Wikimedia Commons
www.Commons.Wikimedia.org

Many thanks to all the incredible photographers, artists,
researchers, and archivists who share their great work.

PLEASE NOTE :
As with all reprinted books of this age that are intended to perfectly reproduce the original edition, considerable pains and effort had to be undertaken to correct fading and sometimes outright damage to existing proofs of this title. At times, this task can be quite monumental, requiring an almost total rebuilding of some pages from digital proofs of multiple copies. Despite this, imperfections still sometimes exist in the final proof and may detract slightly from the visual appearance of the text.

DISCLAIMER :
Due to the age of this book, some methods or practices may have been deemed unsafe or unacceptable in the interim years. In utilizing the information herein, you do so at your own risk. We republish antiquarian books without judgment or revisionism, solely for their historical and cultural importance, and for educational purposes.

Self Reliance Books

Get more historic titles on animal and stock breeding, gardening and old fashioned skills by visiting us at:

http://selfreliancebooks.blogspot.com/

INTRODUCTION

I am very pleased to present to you another great poultry publication – **Domestic Fowl and Ornamental Poultry**. It was written by H.D. Richardson in 1856, making it over 160 years old.

An expert book covering many topics on raising poultry including *Domestic Fowl, Poultry Houses, How to Feed Poultry, The Origin of Our Domestic Fowl, Selection of Stock, Selection of Eggs for Hatching, Web-Footed Birds, The Turkey, Diseases of Fowl, The Duck, Caponizing,* and more.

With coverage of everything from selecting a breed and housing, to feeding and disease, this expert old book is a great place to start if you are considering starting a poultry business, or even just raising your own meat and eggs.

Jackson Chambers

State of Jefferson, November 2017

PUBLISHER'S ADVERTISEMENT.

The Publisher, having found the want of small, cheap Books, of acknowledged merit, on the great topics of farming economy, and meeting for those of such a class a constant demand, offers, in this one, a work calculated to fill the void.

The works of RICHARDSON on the Hog, the Horse, the Bee, the Domestic Fowl, and the Pests of the Farm, are popular in England and in America, and, in evidence of their worth, meet with continued sale both there and here. Hitherto they have not been offered to the American public in an American dress; and the Publisher presents in this Reprint, one of the series, adapted to American wants, and trusts that a discerning Public will both buy and read these little Treatises, so admirably adapted to all classes, and fitted by their size for the pocket, and thus readable at the fireside, on the road, and in short everywhere.

C. M. SAXTON,
Agricultural Book Publisher.

INDEX.

Accidents to fowl, how treated, 87
Age, great, of geese, 82.
Animal food, requisite in feeding poultry, 19; proper kinds, best mode of giving, 18, 19, 20.
Apoplexy, 91.
Ashes and Litter, in a poultry-house, 14.
Asthma, 90.
Aylesbury Duck, 86.

Bankiva Fowl, described, 26; originate the Bantam and Turkish fowl.
Bantam. the origin and varieties of, 25; description of, 51.
Barbary Fowl, 45.
Barndoor Fowl, 55.
Bolton Greys, 45.
Breeding Poultry, advantages attending, 7; profits accruing from, 8.
Buckwheat promotes fecundity in poultry, 19.

Cabin, Cottier's, advantage of poultry roosting there in winter, 16.
Call Duck, 87.
Canada Goose, 75.
Capons, fattening of, 96.
Caponising, 92; objects proposed in, ib.; process, ib.; treatment, ib.; performed on hens, 94; on pullets, ib.; precautions, ib.; sometimes performed on turkeys, geese, and ducks, 96.
Chick, formation of the embryo, 33.
Chickens, when hatched, how to tend before they leave the egg, 32; how they manage to break the shell, 34; when they are to be assisted in liberating themselves, 35; nature of their first feeding, 35; of their housing, 35, 36.
Chinese Goose, 77.
Chittagong, the, 37.
Christmas, goose a favorite dish at, 74.
Cochin-China Fowl, the, 38.
Cock, Domestic, separate feeding requisite for, 17; partialities for particular hens, ib.; dislikes of and their causes, 18; common. description of, 21; his history, 22; a sacred bird with the ancients, ib; importance attached to among ourselves from earliest times, ib.; original country of untraceable. 23; various opinions of the original country of, ib.; reasons for deducing his pedigree from the Jago, Sumatra, and Java fowl, 25; selection of a good, 28; pugnacity of, how repressed, 29; apparent feelings of in crowing, ib.; his fondness of a clean and trim plumage, ib.; gigantic, or St Jago fowl, described, 26; the gold-spangled Polish, 42; silver-spangled Polish, ib.; white-crested black Polish, 43; Dutch everyday layer, 44; Dorking, 45; the Malay, 37; the Cochin-China, 38; the Spanish, 40; cock-fighting among the ancient Greeks and Romans, 22; its introduction into Britain, 22.
Columbian Fowl, 42.
Consumption, 91.
Corns, 92.
Costiveness, 92.
Cramming of fowl, practised by the ancient Romans, 22; cruel and unwholesome practice with geese, 113.

Diarrhœa, 91.
Dietary, varied required for fowl, 20.
Diseases of fowl, 87.
Dorking Fowl, 46.
Dorking Hens good sitters, 37.
Duck, the species admit of a natural threefold division, 83; power to find their food, 84; value of domesticated, ib.; habits of the whole race, ib.; the domestic, ib.; feeding and fattening, 85; hatching, ib.
Dunghill Fowl, the, 55.
Dutch Everyday Layers, 44
Dutch Fowl, the, 55.

East Indian Black Duck, 87.
Egg, shape of, indicates the gender of the future chick, 32.
Eggs, for hatching, how to preserve, 31; for hatching, how to select, 32; during the process of hatching, broken to

INDEX.

show the mode of furnishing the nutriment, 33.
Every-day layers, 44.

Fattening of chickens, 21; of geese, 80; of ducks, 84; of turkeys, 66.
Feeding, separate, advisable for poultry in certain cases, 17; precarious in the farm-yard not to be depended on, 18; substances that may be used in, ib.; necessity of animal food in, 19; mode of giving it, ib.; peculiar, calculated to promote fecundity, how given, ib.; varied, necessary, 20; stimulating, a favorite with fowl, ib.; fattening, 21.
Fever, 91.
Fowl, Domestic, housing for, 9; methods of feeding, 17; the origin of, 21; selection of stock of, 27; hatching of, 32; management of the young brood of, 35; varieties of, 36.
Frizzled Fowl, the, 54.

Game Fowl, 48; management of breeding, 29, 49; varieties of, 50.
Goose, the, 73; common wild or gray-lag, 74; fattening of, 82; the white-fronted, 74; the Canada, 75; the domestic, ib.; the Toulouse, ib.; the Chinese, 77; varieties of the domestic, 78; breeding the, 81; food of the young, 80; keeping, ib.; fattening, 81; French process of fattening, 82; Polish process, ib.; plucking alive for the feathers, 83.
Gout, 92.
Grass-plot, requisite in poultry-house, 14.
Grey-lag Goose, 74.
Guinea Hen, the, 68; its origin, ib.; its characteristics, 69.

Hamburgh Fowl, 44.
Hatching, best situation for, 10; the nest for must be clean, 11; how conducted where there is more than one breed of fowl, ib.; choice of a good domestic hen for, 30; marks of a hen's anxiety for, ib.; how to induce the desire for, ib.; inconstancy of a hen in, how remedied, ib.; over-constancy, how treated, 31; breaking the eggs in, how remedied, ib.
Hemp-seed recommended for increasing fecundity in poultry, 19.
Hen, Domestic, described, 21; the number of hens to be allotted to one cock, 27; selection of a good cock, 28; choice of a good one for incubation, 30; the Malay, a valuable cross-breed, 37; the Cochin-China, 38; the Spanish, 41; the gold-spangled Polish, 42; white-crested black Polish, 43; Dutch every-day layer, 44; Dorking, 47.
Hen-coop, description of a, 16, 17.

Incubation, period of, in the various domestic fowl, 36.

Indigestion, 91.
Inflammation, 89; of the lungs, 90; of the heart, ib.

Java Fowl, 37.
Jumper, the, 53.
Jungle Fowl, description of, 23.

Litter, how pleasing to poultry, 14.

Malay Fowl, 37; a cross from them deserving the breeder's attention, 38.
Moulting, 87.
Muscovy Duck, 86.

Negro Fowl, 55.
Nests for poultry, how made and disposed, 10; those that are most easily cleaned, 15.

Parasites in fowls, 91.
Peacock, the, 60.
Pepper, a favorite relish with domestic fowl, 20.
Perch for poultry, the best, 9.
Pheasant Fowl, 51.
Pheasant, Malay, ib.
Pintado, the, 68.
Pip, 88.
Polish Fowl, 42; the spangled, 43; the white-crested black, ib.; the white, 43.
Pouch, abdominal, of the Toulouse goose, 76.
Poultry, separate feeding of, in certain cases, 17; their dispositions to be observed, 29.
Poultry houses, 9; how to be well kept, 10; how to be warmed, ib.; cleanliness and space for exercise essential to, 12; separate cribs for the diseased requisite in, ib.; separate pens requisite in, 13; ground-plan for, ib.; the house itself described, 12; various requisites for, 14, 15.
Prices of superior poultry, 8.
Profit, of rearing turkeys, 65.
Pugnacity in the cock, how repressed, 29.
Pulse, sorts unwholesome to turkeys, 67.

Rouen Duck, 86.
Rumpkin, the, 53.
Russian Fowl, 54.

Sand, for a poultry house, 14, 20.
Shakebag, the, 86; his origin, ib.
Siberian Fowl, the, 54.
Silky Fowl, the, 53.
Sir John Sebright's Fowl, 52.
Sitting, inconstancy in, how remedied, 30; overconstancy in, how treated in a hen, 31; how to preserve eggs for, ib.; to select eggs for, 32; management of the eggs during the, ib.; when and how to aid, 34; table giving time of sitting and the number of eggs hatchable by the various domestic fowl, 36.

INDEX.

Spangled Fowl, their varieties, 44; confusion in distinguishing them, ib.
Spanish Fowl, 41.
Sussex Fowl, 47.
Swan, the, 70; the mute, 71; the domestic, ib.; the black, 73.

Toulouse Goose, 75.
Turkey, the, 56; mistake of Linnæus in his name for, ib.; original country of, ib.; his introduction into England, 57; origin of his English name, ib.; the wild, 58; his movements, 59; experiments in crossing with the domestic, 60; the domestic, 61; varieties of, 62; best mode of keeping, ib.; treatment of the chickens, 64; feeding, ib.; fattening, 66; the weight of, 67.
Turkish Fowl, the, 58.

Vermin, approved method of ridding poultry of, 14.

Web-footed Fowl, 70.

Yard, an outer and inner, to be attached to a poultry-house, 12.

DOMESTIC FOWL.

*"How grateful 'tis to wake
While raves the midnight storm, and hear the sound
Of busy grinders at the well-filled rack;
Or flapping wing or crow of chanticleer,
Long ere the lingering morn; or bouncing flails
That tell the dawn is near! Pleasant the path
By sunny garden wall, when all the fields
Are chill and comfortless; or barn-yard snug,
Where flocking birds, of various plume and chirp
Discordant, cluster on the leaning stack
From whence the thresher draws the rustling sheaves"*

CHAPTER I.

VIEW OF THE IMPORTANCE OF THE SUBJECT.

POULTRY KEEPING is an amusement in which every body may indulge. The space needed is not great, the cost of food for a few head, insignificant, and the luxury of fresh eggs or home-fatted chickens and ducks not to be despised. In a large collection of poultry may be read the geography and progress of the commerce of the world. The Peacock represents India; the Golden Pheasant and a tribe of Ducks, China; the Turkey, pride of the yard and the table, America; the Black Swan, rival of the snowy monarch of the lakes, reminds us of Australian discoveries; while Canada and Egypt have each their Goose. The large fat white Ducks—models of what a duck should be—are English, while the shining green black ones come from Buenos Ayres. And when we turn to the fowl varieties, Spain and Hamburgh, Poland and Cochin China, Friesland and Bantam, Java and Negroland, beside Surrey, Sussex, Kent, Suffolk, and Lancashire, have each a cock to crow for them.

But we may derive other useful lessons besides those of geography and commerce from the poultry yard. The same principles, the same close attention to food, warmth, and symmetry of form, which have produced perfection in short-horned cattle, Leicester sheep, and thorough-bred horses, have, in a minor degree, afforded us Bantams, "true to a feather," as well as size and beauty in Spanish, Dorking, and Poland Fowl.

Whether poultry keeping can be rendered profitable, is a question which depends on a variety of circumstances, which cannot be alike in two localities; because they depend on the cost of food, and the nett price which can be obtained for the produce in eggs or birds; thus, one person with the free run of a fine dry upland warm common, with a ready market near, may make an excellent profit; while another, bestowing equal care, but confined to a small field of cold soil, may lose nine out of ten of the most valuable young poultry.

Poultry may be converted into money either while living or when dead; or they may be bred, partly for the market, and partly with a view to the disposal of their eggs.

First, as to the profit arising from the disposal of superfluous stock. This depends, of course, in a great measure, upon the quality and character of the birds kept, and hence the breeder should confine his fancy to the more valuable varieties. The expense of feeding and rearing a valuable fowl will not be found to exceed that required for a comparatively worthless one; at least, if at all, only as regards comfort and warmth, which, if properly procured, are not very costly. Poultry of very superior quality and fashionable varieties, especially such as have obtained prizes at any of the first-rate agricultural exhibitions, will fetch a high price. Prize fowl, of extraordinary excellence, bring double price frequently; but of course this is a price given for the *breed*, and not for killing. In all these cases the producer must, of course, allow a fair profit to the *dealer;* he cannot, therefore, reckon on more than two-thirds of the price, yet this will amply remunerate him.

But although there may be doubts about the profits, there can be none about the amusement to be derived from a well chosen collection of domestic birds, and, whether for profit or amusement, the rules to insure success are the same. It will be my endeavor to lay these down as plainly as possible.

Certainly the present, if any, is the time for making a profit by poultry, since all the inferior kinds of grain are cheap and likely to be cheaper. The demand for poultry increases rather than diminishes, and railroads have opened up cheap conveyances to market. The fact is that the great drawback on poultry rearing arises from loss by disease; while the greatest profits are derived from successfully rearing the birds which are most subject to disease at inclement periods of the year.

Ducks and geese are more easily raised than fowl, turkeys, or guineafowl, if there be conveniences of grass and water; but then fine turkeys and fat young guinea-fowl in due season are sure of a sale at a good price. With respect to the poultry of farm houses, which are fed on what would otherwise be wasted or what is collected by the industry of children;—warmly housed, they often thrive better and prove more prolific than the expersively tended inmates of ornamental poultry houses.

In the following pages the most esteemed varieties of poultry and waterfowl will be described. The poultry keeper will find it to his advantage to keep a good breed in preference to a bad one. Some of the more beautiful or valuable kinds of poultry are too delicate to prove profitable; but size, early maturity, and prolific hens, will, under the most unfavorable circumstances, be of more advantage than small, ugly, rarely-laying birds.

CHAPTER II.

POULTRY HOUSES.

BEFORE purchasing your poultry have your house all ready to receive them, or you may do your stock more harm in a few days, by close cooping or cold roosting them, than you can repair in a year. I design showing here how very readily, and at how small a cost, a sufficiently good, and in every respect suitable poultry-house may be erected. I cannot, of course, desire to recommend any restrictions to those whom Providence has favored with wealth. There exists no reasonable objection to such as can afford it gratifying their taste, either as to extent of accommodation or elegance of structure. The poor man, on the other hand, need not lay out one penny, and still may be as successful in his operations as his more wealthy neighbor. It is my object to write for all classes. I shall, accordingly, describe several sorts of poultry-houses, from that on the most perfect and extended scale, to that which can only boast of *barely* answering the purposes for which it is designed.

In nine cases out of ten some outhouse is appropriated to the purpose, without preparation or alteration. But, if consistent with your means, by all means build a proper house. Choose a piece of gravelly soil well drained on a slight declivity, near trees which will afford shade and shelter from winds. The building should be lofty enough to admit the poultry keeper without stooping, because, if it be inconvenient to enter, the chances are that regular cleaning will be neglected. Let the roof be kept weather tight. If slates or tiles are employed, the house should be ceiled in order to protect the fowls from draughts and rapid variations of temperature; in default of lath and plaster a piece of asphelted felt closely nailed makes a cheap and efficient ceiling.

The best perch is made in the shape of a broad double ladder, stretched out so as to form a wide angle; the bars being placed so far apart that one fowl shall not overhang another. If roosting bars be used across the fowl-house, care should be taken that a convenient hen-ladder is always attached to them, and that they are not placed too high. Heavy fowl are apt to break their breast bones in trying to fly down from high perches.

The careful poultry keeper should take a view of the fowl at night after they have gone to roost, and see that they are all comfortable, not too crowded, with room enough for the weak ones to get away from the strong, who are apt to tyrannize. The floor must be dry, and covered with fine gravel or sand, and it should be swept clean every day. Nothing injures the health of fowl more than bad smells. To obviate this always keep a basket of slacked lime or old mortar in a corner with a shovel, so that you may shake some over any dirt. The sweepings, if kept quite dry, form most valuable manure. For the same reason have the interior walls frequently whitewashed, and the window open in fine weather. If the window can be filled with a venetian blind so much the better. The door should have a hole at the bottom with a sliding panel to admit the poultry during the day—by keeping it locked you have a better chance of gathering plenty of eggs. If you have no windows, movable loose boards fitted to the door may be useful to admit air.

As warmth is so requisite to poultry, it will be an advantage if one side of the poultry house be against the outside wall of a kitchen or boiler-house, or a hot water pipe running through it from the hot-house will well repay the outlay. With a sweet clean warm poultry house you will have plenty of eggs long before more careless neighbors.

As to the nests, the great point is that they should be near to the ground, easily cleaned, and not too large. If they are too large two fowls will often try to sit in the same rest at the same time. If there is any difficulty in getting into them, hens are apt to drop their eggs on the ground. Nests may be made of wood, or basket-work; there should be a small ledge to prevent the eggs from rolling out. A little old mortar or wood ashes laid at the bottom will tend to keep the nests clean. Straw and hay both make good lining for nests.

If the nests are arranged in two stories there should be a broad ledge wide enough for a hen to walk on in front of the top row, like the platform of a drawing-room verandah, and a hen-ladder should be placed at each end, but nests are better on the ground.

It is very advantageous to place fowl which are sitting in a retired situation where they will not be annoyed by other fowl, and where, when the hatching takes place, they can be cooped with their young out of danger, with a dry yard or close cropped lawn in front to run on. Many hens, as well as peafowl and turkeys, are vicious, and will try to destroy a rival brood.

A small box, about a yard square, with a hard dry floor, and a movable

POULTRY HOUSES.

wooden top, is excellent as a *sitting*-room for hens. I have seen an old cucumber frame covered with wooden slabs successfully arranged for bringing up early broods.

Be sure before you put a fowl to set that the nest is perfectly clean; if the hen becomes infested with vermin she pines and cannot set close.

It will often be found cheaper to make a good fowl-house at first, than to be continually adding and patching.

Of course if you have more than one breed of fowl, they must be kept separate, if you intend to keep the race pure. Where this is attempted, an enclosure adjoining the poultry-house, with three divisions of iron wire, will be found useful, if the space and cost can be spared. In these enclosures in wet cold weather, the poultry can be confined, with room to scratch and feed. The largest division will be for your laying hens and turkeys, and miscellaneous stock. In this space you can muster them, accustom them to be fed, and see that all are in health, and make the close observations which are needful for success. In the second you can place hens with young broods before they are strong enough to mix with the other fowl. In the third, and smallest, poultry for fatting. If just large enough for them to enjoy the air

POULTRY HOUSE.

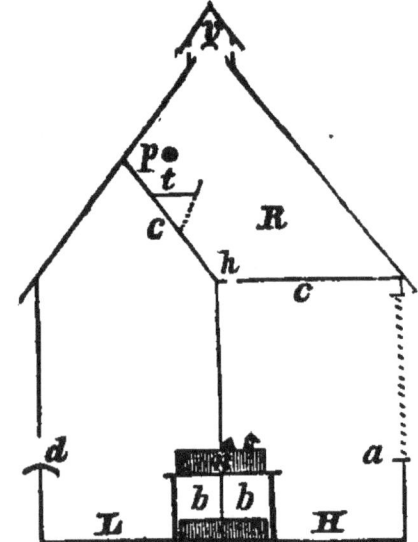

TRANSVERSE SECTION.

L, the laying apartment; *H*, hatching-room; *n*, nest boxes; *b, b*, boxes and troughs for water, feed, sand, &c.; *d*, an opening, or door, for entrance and going out; *a*, latticed window, for ventilation; *R*, roosting-place, separated from apartments below by the ceilings *c, c*; *h*, ventilating hole; *v*, ventilator on roof; *p*, roosting pole; *t*, trough for dung.

without being able to run about much, with shade, sun, plenty of clean water, and food, they will generally thrive better than when cooped. A few good coops, either of wood and wire, or wicker, with the top thatched, should always be at hand. These should be made so as to shut up the chickens if necessary, as well as the hens. If the fowl-house is large enough have a small sink in one corner where it is light, and if it is not large enough, put in the yard, under shade, a large glazed earthenware pan, and fill it with fine sand, or ashes, or slacked lime, or burned oyster shells, as a dust bath for the fowl. By placing the stuff in a pan it is easily changed from time to time.

If you are obliged to put up with a small lean-to or other confined place, for your fowl-house, at any rate take care to keep it clean, for warmth, cleanliness, and judicious feeding are the cardinal maxims for poultry management.

Nothing more is necessary for the keeping poultry with profit and advantage, beyond having a small shed or light building, formed in some warm, sunny, and at the same time, sheltered situation, fitted up with proper divisions, boxes, lockers, or other contrivances for the dwelling of the different sorts of birds, and places for their laying in. This and *cleanliness* suffice.

"Cleanliness," says Mr. Beatson, "with as free a circulation as possible, and a proper space for the poultry to run in, is essential to the rearing of this sort of stock with the greatest advantage and success, as in narrow and confined situations they are never found to answer well."

In every establishment for poultry rearing there ought to be some separate crib or cribs, into which to remove fowl when laboring under disease; for not only are many of the diseases to which poultry are liable highly contagious, but the sick birds are also regarded with dislike by such as are in health; and the latter will generally attack and maltreat them, thus at the very least aggravating the sufferings of the afflicted fowl, even if they do not actually deprive them of life. The moment, therefore, that a bird is perceived to droop or appear to be pining, it should be removed to one of these infirmaries.

Separate pens are also necessary to avoid quarreling among some of the highly-blooded breeds, more particularly the game fowl. They are also necessary when different varieties are kept, in order to avoid improper or undesired commixture from accidental crossing. These lodgings may be most readily constructed in rows, parallel to each other; the partitions may be formed of lattice work—they will be rather ornamental than otherwise, and the cost of their erection will be but trifling. Each of these lodgings should be divided into two compartments, one somewhat larger than the other. One compartment is to be close and warm for the sleeping-room; the other, and the larger one, should *be airy* and open, that the birds may enjoy themselves in the day-time; both should be kept particularly dry and clean, and be well protected from the weather.

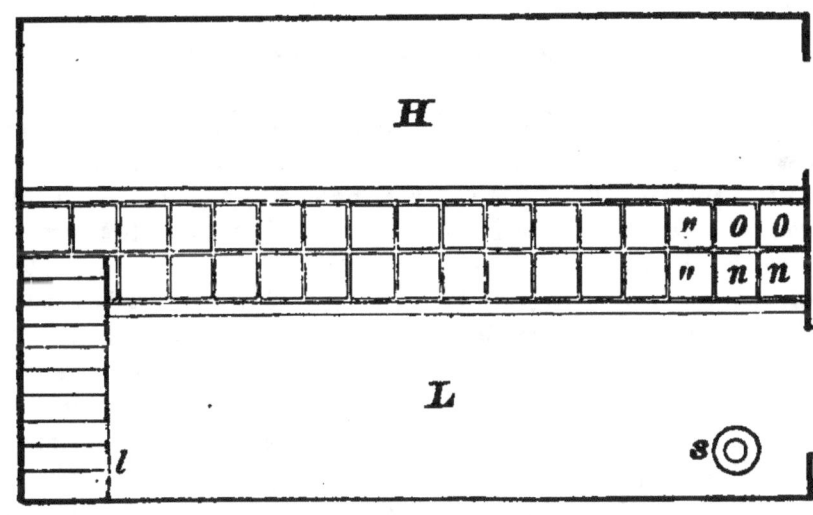

GROUND PLAN.

L, laying-room; *H*, hatching-room; *n, n*, nests for laying; *o, o*, nests for hatching; , ladder; *s*, stove.

Attached to the house should be a well-drained yard, with a division of wire or trellis work for every ward, with water in each; and it will be advantageous to have outside of this yard a wider range of turf and gravel, where the fowl can be more at large. When different broods are kept, and it is desired to keep them apart, the larger yard must be shared in turn by the inhabitants of the different wards. The hatching-ward and the feeding-ward should be kept separate. A roosting and hatching-ward for ducks and geese, with a small pond, accessible to all the inhabitants of the poultry yard, should be added.

WATER DISH AND TANK.

Every poultry-house should be provided with a sufficient quantity of small sand; or, if such cannot be procured, clean ashes are a good substitute; pieces of chalk are also a useful, nay, necessary adjunct; *crude lime* acts, however, as a poison. Some horse-dung or chaff, with a little corn through it, is also a source of much amusement to the birds; and it should be borne in mind that *amusement*, even in the poultry-yard, is materially conducive to health. The ashes and litter should be frequently changed, and had better also be kept in little *trenches*, in order that they may not be scattered about, giving a dirty or untidy appearance to the yard. When, however, your fowl have a run in a garden or field, of average extent, this *artificial* care will be replaced by nature.

If the court be not supplied with a little grass-plot, a few squares of fresh grass sods should be placed in it, and changed every three or four days. If the court be too open, some bushes or shrubs will be found useful in affording shelter from the too perpendicular beams of the noon-day sun, and probably in occasionally screening the chicken from the rapacious glance of the kite or raven. If access to the sleeping-room be, as it ought, denied during the day, the fowl should have some shed or other covering, beneath which they can run in case of rain: this is what is termed "*a storm house;*" and, lastly, there should be a constant supply of *pure, fresh water.*

Fowl frequently suffer much annoyance from the presence of vermin, and a hen will often quit her nest, when sitting, in order to get rid of them. This is one of the uses of the *sand* or *dust* bath; but a better remedy, and

POULTRY HOUSES. 15

RUSTIC POULTRY-HOUSE.

one of far speedier and more certain efficacy, is to have the laying nests composed of dry heather and small branches of hawthorn, covered over with white lichen. These materials, rubbed together by the pressure and motion of the hen, emit a large powder, which, making its way between the feathers to the skin, is found to have the effect of dislodging every sort of troublesome parasite.

The fowl-house should also be frequently and thoroughly cleaned out, and it is better that the nests be not fixtures, but formed in little flat wicker baskets, like sieves, which can be frequently taken down, the soiled straw

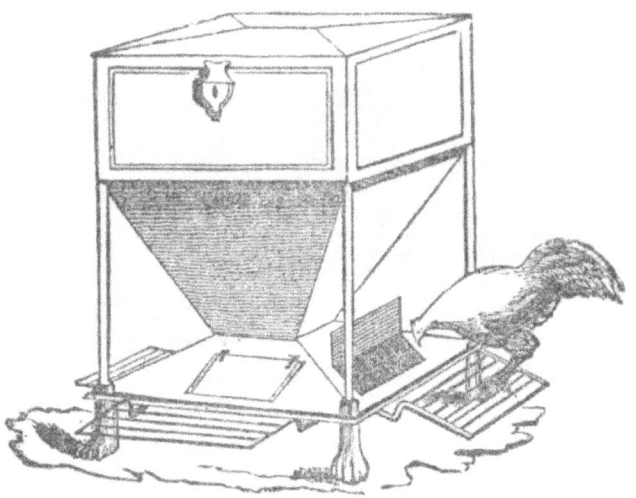

FEEDING-BOX.

thrown out, and themselves thoroughly washed: or formed of wooden boxes, with a sliding bottom. In either case, hay is objectionable, as tending to the production of these vermin. Fumigation, at no very remote intervals, is also highly to be commended. Nothing is of more importance to the well-being of your poultry than a good, airy walk. These maxims cannot be too often impressed on the poultry keeper.

A COTTAGE POULTRY-HOUSE.

As good a mode of rearing fowl as can be adopted is the old custom of suffering them to roost on the rafters of the room in which the cottier keeps his fire; and it is, perhaps, owing to the warmth thus afforded to the birds, that, during winter, when eggs are scarce, and consequently at a high price, they will be procurable from the humble cabin, when they have long vanished from the elaborately constructed, but less *warm* poultry-house of the more affluent fancier.

Should circumstances, however, render the keeping poultry in the cabin objectionable or unadvisable, a very sufficient place may be erected for them against the outside of the cabin wall; and, if possible, the part of the wal against which the little hut is erected should be that opposite to the fireplace within—thus securing the necessary warmth. If shelter be required, it can be obtained by means of a few bushes, or a wall of sods; the neighboring roads will serve as an ample walk; the nearest stream will slake their thirst. A few laying-nests may be placed in a warm corner of the cabin, and the poultry of the poor cottier will thrive as well, and yield great a profit, as those kept in the best appointed establishments.

COOP AND FEEDING-BOX.

By confining a hen some hours in the day to the coop, she is prevented from rambling into danger, and yet has the liberty of enjoying fresh air, and the pleasure of seeing her chicks run in and out through the bars, returning to her when her voice warns them to seek shelter with her in the friendly coop, on the approach of a shower or of any other danger. The instincts of the young birds will generally lead them to obey her voice, even though it be that of a step-mother.

TENT-SHAPED COOPS.

CHAPTER III.

HOW TO FEED POULTRY.

Do not feed your hens too highly before they begin to lay, or while laying, or immediately after ceasing to lay, unless you wish to fatten them for table use; for as soon as a fowl begins to fatten she stops laying. You must, therefore, separate the two classes of fowl, layers and fatteners, at all events at feeding time. Make some separate provision for your cocks; if they are only fed in company with the hens, they are apt to think too much of their mistresses and to neglect their own appetites; and recollect that to have *strong* chickens, you must have strong cocks, which an ill-fed bird cannot be expected to prove.

You should also make separate provision for such fowl as are bullied or oppressed by the rest. Fowl are much given to *jealousy;* the *cock's favor* is sometimes the cause of this, but by no means invariably so, and, indeed, the cause is not at all times to be ascertained; however obscure the *cause*, it is incumbent on the poultry fancier to prevent the *effect*, by adopting the separative system at the times I have indicated. I myself have met with instances of a cock forming a partiality for a particular hen. In such an occurrence, which is easily recognizable by the cock's continually running at that particular bird, to the neglect, or comparative neglect of the others, it is better to remove the favorite *at once;* if you do not do so, quarrels will ensue; this hen will nearly always be made a victim, and in many cases the quarrels on her account will give rise to other and more general affrays. On such occasions the cock usually interferes and endeavors to establish peace; he almost invariably does so when the contest is carried on *per duello* when, however, a number of his mistresses fall upon one, his interference is of little avail; and, as if he were conscious of this, in such cases he usually leaves the poor favorite to her fate. I have also known a cock to take a *dislike* to a particular hen. This is a much more rare case than the preceding, and I have no doubt of its cause; it is this: when a vigorous, healthy cock is mated with *very few* hens, he is very persevering in his attentions to them: when hens are in *moult* they will not accept of any such attentions. In most instances of this kind that have fallen under my own observation, I have found the hen thus victimized by her lord to have been moulting, and to have incurred his hatred by a refusal of conjugal rights. The cock will sometimes fall upon a hen newly introduced into your yard, especially if of a different color from his other mates.

Fowl about a farm-yard can usually pick up a portion of their subsistence, and that probably the largest portion, and, of course, in such situations poultry-keeping decidedly *pays best*. I must, however, particularly caution my readers against *depending* for the support, even of their *non-fattening* poultry, wholly upon such precarious resources, and I shall, accordingly, proceed in my advice as if no such resources existed.

The substances that may be used in poultry feeding are very numerous and various—cabbage, turnips, carrots, parsnips, mangelwurzel, oats, wheat, barley, rye, indian corn, and other grains, substances too well known to require, and too numerous to be worth the trouble of enumerating. It will not answer to feed fowl wholly upon *any one variety* of food; neither will it be found advisable to feed wholly upon any one *class* of food. I must speak of the latter point first. Fowl require a mixture of *green* food with *hard* food, fully as much as horses or cattle do. When the birds have the advantage of an extensive walk, they will find this for themselves; when they do not possess such an advantage, you must provide green food for them. Some do so by providing the birds with cabbages or other greens *chopped small*. My plan is to *fasten* heads of cabbages, lettuce, rape, or other green herbs, to some fixture, by means of their roots, and to let the fowl peck for themselves. This practice not merely prevents waste, but is, in consequence of the *amusement* it affords, decidedly conducive to health. When you find it difficult to obtain green food, you will find that *turnips* will answer equally well. To prepare these they should be sliced one way, and then sliced across, so as to be cut into small *dice*. This is *troublesome*—granted; but no man deserves to have a good stock of poultry, or anything else, if he declines taking trouble. If it be necessary to employ hired labor for the purpose, the stock must be very large, and will unquestionably pay. The same yellow turnips, boiled soft, and mixed with bran or pollard, or given by themselves, are also capital feeding, especially for a *change*. Carrots and parsnips may be used for this purpose, prepared in either of the modes recommended for yellow turnips. Of *mangel*, as food for poultry, I cannot say much, valuable though it decidedly is for other purposes; the birds do not generally like it, and I have found that, even where they do eat it, it does anything but promote their laying; oats are useful as forming a *portion* of fowl's feeding; but it will not answer for keeping them upon altogether; the *hulls* are very indigestible, and this food is, besides, of too *stimulating* a nature; yet a few handsful are well spent on your fowl. When *damaged* wheat can be bought at a low price, it may be used for the feeding of poultry with much profit and advantage; when no such thing can be procured, however, and when it is proposed to feed them upon the sound, marketable article, it will not pay. The same may be said of *barley*, which is also objectionable as acting in a *purgative* manner—it is useful as an occasional feed, when fowl are over fed. *Rye* is usually a cheaper description of grain than any other, and *damaged rye* may be used, to a limited extent, with impunity, even when affected with the *ergot* which exercises so powerful an influence upon the systems of all female animals possessing a uterus. As this same *ergot*, however, is frequently the cause of severe illness when human beings happen to eat bread made of rye tainted with it, poultry should not be suffered to eat *too freely* of it. Indian corn is a most capital food both for store and fattening fowl, and may be used in larger quantities than any other.

The sweepings of grain warehouses, consisting of all kinds of grain, may freequently be purchased on cheap terms, and are well suited for poultry, but, if given to fowl, the peas and beans must be sifted out.

One circumstance connected with the feeding of poultry, and that a most important one, is not sufficiently well known—I allude to the necessity they are under of obtaining *animal food*. Of course, when the birds possess the advantage of an extensive run, they can themselves peck up insects, worms, snails, or slugs; and as in the case of ducks, &c., frogs and other small reptiles; but in cases where they do not possess this advantage, it is necessary that you cater for them. I have always experienced the best effects, especially as manifested in greatly increased *laying*, of giving scraps of animal food about twice or thrice a-week to the fowl; the best mode is throwing down a bullock's liver, leaving it with them and permitting them to peck at it *ad libitum*. This is better in a raw than in a cooked state.

In winter, in order to supply the place of the insects and other animal food they can pick up in summer, I give them once a week fat meat, together with any meat bones to peck, and also barley made hot in a saucepan without water and given warm. Hot potatoes are always good food, small potatoes may be picked out, and steamed for the purpose, if you keep a garden. But meat is indispensable, if you wish to have eggs in winter.

Several substances have been at different times recommended as calculated to increase the fecundity of the various classes of the feathered inhabitants of the farm yard, amongst these, perhaps, hempseed and buckwheat are pre-eminent. There can exist no doubt of the peculiar efficacy of these seeds in this respect when *properly used*, but neither can it be denied that in some cases this *objectionableness* is undoubted.

When a hen *pines*, or seems disposed to be *thin*, you need not hesitate in giving buckwheat with even a liberal hand; but you must so manage as not to permit such hens as are disposed to become *too fat* to share in this department of your bounty. According as hens take on fat they usually fall off in laying, and this should be particularly kept in mind in feeding. When hens are disposed to flesh, you will find *hempseed* the best promoter of laying; at the same time it will be necessary that you restrict them as respects other descriptions of food, fattening and laying being nearly always incompatible with each other.

Fowl of all kinds require *sand, gravel*, as an aid to digestion, being, in fact, necessary to promote *trituration* in the gizzard, as well as to supply calcareous matter for their egg-shells. You should, therefore, always have a supply placed within their reach. This, I must admit, applies more immediately to such fowl as are kept in a confined yard; when the walk is at all extensive, the birds can usually peck up enough for themselves. Freshwater gravel is the best; and if you live near the sea, and wish to use sand so easily obtainable from the beach, you should first wash it in two or three waters. Where no sand of any kind can be obtained, as in towns, you can use chalk, bruised oyster shells, or freestone; if the latter, you had better wash it well first: you will, of course, pound before placing it in the yard.

I have observed that fowl require a varied dietary.

In the morning, about seven o'clock, in spring and autumn, but at six in summer, let the fowl out, and permit them to roam about till nine, when give grain, to the amount of about a handful to every three birds; they will then amuse themselves about the place, during which time they will peck up a good deal; about three o'clock feed them again on grain to about the same amount, besides which give whatever potato, turnip, or other refuse is going. The liver should lie in the yard, and they can get green feed for themselves. In winter the affair assumes another aspect; all feeding, but more particularly the grain, must be greatly increased in quantity. As you now cannot procure green food, or at least can only do so with difficulty, and at an expense that will seldom pay, you should resort to the chopped turnips.

Cayenne pepper, indeed, all descriptions of pepper, especially the cayenne in pods, will be found a favorite with fowl, and will be greedily devoured by them; it acts as a powerful stimulant and remarkably promotes laying, and, when mixed in a ground state, with boiled meal, will be found productive of the best effects. In this, however, as in every thing else, let moderation be your ruling principle.

A different system should be adopted in treating poultry for the table, and for the laying and breeding department. The great secret of having fat chickens, is never to let them be thin. But, to fatten, you may either enclose them in a small space, or absolutely coop them up. Coops should be placed in a warm—rather dark place; be high and large enough for each fowl to be comfortable without moving about, not more than three fowl in each division, so that they can see without touching each other; the back part of the floor should be grated to allow the dung to fall through, and this must be removed every morning. The troughs are generally made too low; they would be better, raised an inch; and, instead of wood, should be of coarse pottery or glass, both of which are very cheap, and can be easily kept clean. Starve the fowl for a few hours after cooping, and then supply them frequently, and at regular intervals, with as much food as they will eat, and no more, clearing the trough each time after they have fed. Rice boiled will be found very fattening; and by a constant variety of food, the fowl will be induced to eat, and ought to be quite fat in a fortnight.

But above all, it must be remembered, that to do any good, chickens put up for fattening, require regular attention, and at stated hours.

CHAPTER IV.

THE ORIGIN OF OUR DOMESTIC FOWL.

THE Domestic Fowl, styled by zoologists *Gallinæ*, from the Latin word *gallus*, a cock—is distinguished by having the crown of the head usually naked and the skin raised in a fleshy protuberance, called a comb—a protuberance varying in size and form in different varieties. The base of the lower mandible (beak) is likewise furnished with fleshy, lobular appendages, called *wattles*; the tail is carried erect, and is composed of two planes folded together at acute angles. In the male, the central feathers of the tail are elongated, and fall gracefully over the others. The feathers of the neck are ample in quantity, are either long and hackled or short and truncated. The plumage of the male bird is characterized by considerable brilliancy and beauty; that of the female is unobtrusive, matronly, and comparatively dull. The cock tribe is extremely hardy, and endures all changes of temperature and climate with impunity, as is proved by these birds being found to exist in nearly every country of the world, *from* the warmest to the coldest zone.

The domestic cock appears to have been known to man from the very earliest period. Of his real origin little appears to be known, and the period or manner of his first introduction into Greece, or southern Europe, is involved in the greatest obscurity. The cock has certainly ever held a prominent position among birds; he occupied a conspicuous place at the shows of the Greeks and Romans in the days of old; his effigy was engraved, and is still to be seen upon many of the medals and coins; and he has been expressly dedicated to several of their favorite deities—as Apollo, Mercury, Mars, and Æsculapius.

At a Roman banquet this bird formed a principal dish, and poultry were even then carefully reared and fattened, as well as *crammed*. Nor was the pugnacious disposition of the cock even then unknown, or lost sight of, as a means of amusing man; for cock-fighting was seriously entertained and encouraged as at once a religious and a political ceremony. The islands of Rhodes and Delos are said to have furnished the fattest birds for the table, as well as the most enduring and unflinching champions of the ancient cock-pit.

It is strange that a practice so barbarous as that of cock-fighting should owe its origin to classic times, and to one of the most learned and enlightened nations of antiquity—the GREEKS. It was introduced into these islands by the Romans, and it was, perhaps, the occasion of making us acquainted with the domestic fowl. For a long period cock-fighting was practiced in England as a royal pastime, and exhibited as such before public assemblies with pomp and show, and it continued to be sanctioned, both by law and custom, until about 1730. Up to this time it was—I suppose in allusion to the well-known connection this bird had with St. Peter's denial of our Savior—a favorite amusement at or about Shrovetide, and was even in vogue at public schools, with the express sanction of the schoolmaster, who furnished the boys with cocks for the purpose.

However much the cock has occasionally suffered, he has, on the other hand, to boast of having ever been regarded as a bird of the very highest consequence and respectability. From time immemorial his "shrill clarion" has "ushered in the morn;" and he has likewise had consigned to him the important power of dismissing ghostly visitants to their more appropriate dwelling in the tomb. The ghost of Hamlet's father, about to make a most important disclosure to his loving son, suddenly hears the *crowing* of the *cock*, on which he announces no less abruptly that he "snuffs the morning air," and leaving half his say unsaid, returns incontinent to all the gloomy and unrevealed horrors of his mysterious prison-house. As Shakspeare so beautifully writes, too, the office of cock-crowing is likewise, at a certain season, rendered still more important—

> "Some say that ever against that season comes,
> Wherein our Savior's birth is celebrated,
> The bird of dawning singeth all night long;
> And then, they say, no spirit walks abroad.
> The nights are wholesome—then no planets strike,
> No fairy takes, nor witch has power to harm;
> So hallowed, and so gracious is the time."

As I have already observed, to pronounce with any degree of certainty as to the original country of the domestic cock, or to refer *positively* to *what known* wild species we are to look for his primitive type, would prove a labor equally difficult and presumptuous, the date of his original domestication belonging to so remote a period as to be now wholly lost; but nevertheless, there are races of poultry that, still possessing a wild and apparently truly feral type, would seem to afford the strongest evidence of originality

SONNERAT'S JUNGLE FOWL.

Several authors of the highest respectability and most unquestionable erudition—among whom Buffon and Sonnerat—have endeavored to show that all the varieties of domestic fowl with which we are now acquainted sprang originally from one primitive stock. This opinion has obtained many advocates. Zoologists are, in general, apparently possessed with an anxious desire to curtail, as much as possible, the number of primitive types whence the several races of animals have sprung; with poultry, however, this desire must be frustrated. Dampier saw wild hens at Puloncondar, Timor, and St. Jago. Sonnini describes wild cocks which he saw in the forests of South America. Temminck procured wild cocks from Java, Sumatra, and Ceylon; and all these birds differed essentially, in character and appearance, from all our then known domestic races—from those found by Sonnerat in the Indies—and, finally, from each other. This statement, like many other novelties, though scouted at the time by Sonnerat and others, who, bigoted to their own pre-declared opinion, were, of course, interested in their contradiction, have since been amply and authoritatively confirmed.

I have neither the wish nor the intention to waste my own time, or that of my readers, by entering upon the useless, unsatisfactory, and often interminable paths of controversy.

It has been very generally supposed, and most commonly asserted, that the domestic cock owes his origin to the Jungle fowl of India. I hold that he does not—that he, in fact, differs as much from that bird as one fowl can well differ from another; they will certainly breed together, but so will the hare and rabbit. Read, however, the following description of the Jungle fowl, and, if you can, point out its counterpart among our domestic stock:—

It is about one-third less than our common dunghill cock, being (the comb not included in the measurement) about twelve or fourteen inches in height. The comb is indented, and the wattles certainly bear some slight resemblance to those of our common cock; but the naked parts of the head and

. JAVANESE JUNGLE FOWL.

throat are much more considerable. The feathers of the head and neck are longest on the lowest parts, and differ both in structure and aspect from those of other cocks, whether wild or tame. The Jungle hen is smaller than the cock, has neither comb nor wattles, and the throat is *entirely covered with feathers*—a very remarkable distinction from our domestic hens. The space round the eye is naked, and of a reddish color; the under parts are furnished with plumage, similar to that of the same parts of the cock: but, in addition to these peculiarities, the Jungle cock possesses still another, which, however, the hen does not share with him—viz., the mid-rib, and stem of a portion of the feathers is considerably expanded, forming a white stripe along the whole feather, as far as the tip, where it expands, becomes broader, and forms a gristly plate of a rounded form, whitish, thin, and highly polished; this gristly substance is still more remarkable on the wing feathers than on any other part, the tip, indeed, of the wing feathers forming a less brilliant plate, solid as horn, and as firm and unyielding to the touch. These plates are of a deep red color, and by their union, form a plate of red maroon which looks as if it were varnished.

There are, however, two wild-cocks in whi h we find sufficient points of resemblance to our domestic varieties, and these answer the purpose of terminating our somewhat unsatisfactory search. I allude to the gigantic bird or Jago fowl of Sumatra, and to the diminutive denizen of the wilds of Java. The reasons for supposing these two birds to be the veritable originals of our domestic poultry, may be summed up briefly thus:—

I.—The close resemblance subsisting between their females and our domestic hens.

II.—The size of our domestic cock being intermediate between the two, and alternating in degree, sometimes inclining towards the one, and sometimes towards the other.

III.—The nature of their feathers, and their general aspect, the form and mode of distribution of their barbs being the same as in our domestic fowl.

IV.—In these two birds do we alone find the females provided with a crest and small wattles, characteristics not to be met with in any other known wild species. You will meet with these characteristics in the highly bred Spanish fowl.

Notwithstanding these analogies, however, domestication has so changed the form of the body, and of its fleshy appendages, that we might find it rather a difficult task to refer any modern individual variety to its primitive stock: we must, in order to understand fully the causes that produce this difficulty, recollect the constant, and frequently careless, crossing one bird with another, and the very frequently *promiscuous* intercourse that takes place in a state of domesticity, taking, likewise, into consideration changes of climate, variety of treatment, and numerous other causes.

We cannot, however, find any difficulty in at once recognizing the large and powerfully-limbed bird of Sumatra (called also the Jago fowl), the appropriately styled "Gigantic Cock," or *Gallus giganteus* of zoologists, as the original type to which we owe the Paduan and Sancevarre varieties.

To the more diminutive Bankiva cock, we are, on the other hand, indebted for the smaller varieties, improperly designated Bantams, and the, so-called, Turkish fowl. By crossing, peculiarities of climate, management, &c., have been produced from these:—

I.—The cock with small crest and wattles, furnished, also, with a tuft of feathers, which some writers have supposed to be produced by the juices that ordinarily go to furnish nourishment for the comb taking another form, and developing themselves in the production of the tuft. These approximate most nearly to the original Sumatra stock, and we may recognize their domestic representative in the varieties of the Polish breed.

II.—The ordinary cock, provided with comb and wattles, but no crest c tuft of feathers; this seems the intermediate variety.

III.—Diminutive cocks, ordinarily known as Bantams, with, in some varieties, the tarsi and toes covered with feathers; but this is not invariably the case.

I should here describe the two races to which I have stated it as my opinion, that we are indebted for our domestic varieties.

The wild cock, justly termed the "Gallus giganteus," and called by Marsden the "Jago Fowl," is frequently so tall as to be able to peck crumbs without difficulty from an ordinary dinner-table. The weight is usually from ten to thirteen or fourteen pounds. The comb of both cock and hen is large, crown shaped, often double, and sometimes, but not invariably, with a tufted crest of feathers, which occurs with the greatest frequency, and grows to the largest size in the hen. The voice is strong and very harsh, and the young do not arrive at full plumage until more than half grown.

There was, some years ago, in the Edinburgh Museum of Natural History, a very fine specimen of the Jago fowl; it was said to have been brought direct from Sumatra, and, in most respects, closely resembled the common large varieties of domestic cock. In this specimen the comb extended backwards in a line with the eyes; was thick, slightly raised, and rounded on the top, almost as if it had been cut; the throat bare, and furnished with two small wattles. The neck and throat hackles of a golden reddish color, some of them also springing before the bare space of the throat; the hackles about the rump, and base of the tail, pale reddish yellow, long and pendent; the center of the back, and smaller wing coverts, of a deep chestnut brown, the feathers having the webs disunited; the tail very full, and of a glossy green color. The greater wing coverts of a glossy green, with the secondaries and quills of a faint golden yellow; under parts of a deep, glossy, blackish green, with the base of the feathers a deep chestnut brown, occasionally interrupted, so as to produce a mottled appearance. This bird measured very nearly thirty inches in height, comb included, and making allowance for the shrinking of the skin; the living bird must have been upwards of thirty-two inches high.

The Bankiva fowl is a native of Java, and is characterized by a red, indented comb, red wattles, and ashy-grey legs and feet. The comb of the cock is scolloped, and the tail elevated a little above the rump, the feathers being disposed in the form of tiles or slates; the neck feathers are gold color, long, dependent, and rounded at the tips; the head and neck are of a fawn color; the wing coverts a dusky brown and black; tail and belly black. The color of the hen is a dusky ash-grey and yellow; her comb and wattles much smaller than those of the cock, and, with the exception of the long hackles, she has no feathers on her neck. These fowl are exceedingly wild and inhabit the skirts of woods, forests, and other wild and unfrequented places. These Bankiva fowl are very like our Bantams, and, like those pretty little birds, are also occasionally to be seen feathered to the feet and toes.

CHAPTER V.

SELECTION OF STOCK, AND CHOICE OF COCK AND HENS FOR SITTING.

COLUMELLA is, perhaps, among the earliest authorities we can cite on the subject of the breeding and management of poultry, and he thus delivers himself on a very important subject, viz., the number of hens to be allotted to each individual cock:—

"Twelve hens shall be enough for one good cock, which will cause the progeny to be more of a color; but yet our ancestors used to give only five hens to one cock, thus producing a diversity of color. To have the hens all of one color is preferable, some white, and these are considered the best layers."

Bradly, in his *Farmers' Director*, advises one cock to be left with seven or eight hens, and hints that if a greater number be allowed him, the eggs will not prove fertile. The author of the *Complete Farmer*, and the writer of the article on poultry, in *Rees' Encyclopedia*, recommend the same number.

M. Parmentier, a very eminent French writer, says, that one cock is *much more* than sufficient for fifteen, or even twenty hens, provided he be a young, vigorous and healthy bird.

Those who breed game fowl for combat, and whose object is, of course, the production of strong chickens, limit the number to four, or at most five. Mr. Mowbray says, that in winter, or cold and damp weather, a cock should only have four hens. M. Bose (*Encyclopedia Methodique*) says, that in spring alone should any cock have fewer than twenty hens. M. Dickson says, that the number of hens allowed to one cock should vary with the object you have in view; and Mr. Nolan, a most excellent judge, thinks that in order to secure a prime breed, a cock two years old should not have more than five hens.

If you look for profit to the production of eggs alone, I should say that one cock—if a stout, young, and lively bird—may have as many as twenty-four hens. If, however, you want to obtain strong and thriving chickens, you must restrict him to six, or at most, eight. If your object be the improvement of a worn-out or degenerate breed, the fewer hens you allow to one cock the better, and you should not, at any rate, allow him more than three.

As to the selection of a good cock, Columella thus instructs us:—"It is not good to keep a cock if he be not stout, hot and knavish, and of the same color as the hens are, and with as many claws; but in his body to be higher raised, his comb to be high and red as blood, and straight withal; his eyes black or azure color; his beak short and crooked, with a grey crest, shining like red or white, and all his feathers, from the head to the breast, to be of a changeable color, varying like gold or yellow; his heart large and big; his muscles on his wings big like one's arm, with long wings; his tail fair and long, with two ranks of crooked and rising feathers; and to be oft crowing is a sign of lusty courage. The red color is thought to be the best cock; his legs short and strong, his thighs great and thick, and well covered with feathers, and his legs armed with long spurs, rough and pointed—straight in body, light, fierce, eager in battle, vigilant, ready, and often crowing, and not easily feared."

Markham, in "*Cheap and Good Husbandrie*," almost repeats the directions of Columella *verbatim*, and guarantees their correctness with the authority of his own opinion.

M. Parmentier recommends the cock to be chosen of a middling size; carrying the head high, having a quick, animated look, strong, shrill voice, short bill, very red comb, large wattles, broad breast, strong wings, black or dull red plumage, thighs muscular, spurs strong, claws bent and sharp, free in his action, a frequent crower, and frequently scratching the ground in search of worms, not, however, for himself, but to treat the hens.

Not to weary my readers with an unnecessary citation of too many authorities, I may just observe, for their direction, that the cock should be *in perfect health;* feathers close and rather short, chest compact and firm; full in the girth; lofty and elastic gait; thigh large and firm; beak short, and thick at its insertion.

Next to health and strength, age is to be duly considered. Neither select a cock that is too old, nor one that is too young; let the age be from a year and a half to three years and a half. Some cocks retain their vigor till they are even past six years old, and some make a display of unquestionable virility at the premature age of five or six months. It is far better, however, for the fancier "to be sure than sorry." Secure a young and vigorous bird at the summit of his prime, steer equally clear of premature and often deceptive developments, and of incipient age and decrepitude—avoid all extremes.

Mascall, following Columella and Stephanus, says—"The signs of a good hen are these—a tawny color or a russet are accounted the chiefest colors; and next, those hens which have the pens of their wings blackish, not all black, but partly so. As for the grey and the white hens, they are nothing so profitable."

Markham tells us that we must lay even more stress on the selection of a hen than on the choice of a cock, and insists on "grey, grissel, speckt, or yellowish—black or brown is not amiss."

These directions may have been all very well in olden times, ere the many new and valuable varieties of fowl now known were familiar to the poultry-yard, but as far as color is concerned, they can no longer be followed, unless with respect to the common Dunghill breed. Among these latter you may, of course, make what selection you please as to color, but the more valuable and distinctly marked varieties have each its own hue, and you must, consequently, just take them as you get them. Perhaps the best mode of forming a conclusion as to the most profitable color would be to keep a memorandum-book, and to enter regularly the age, color, and every other particular connected with your hens; and, of course, keep also a correct account of their proceeds, whether as to eggs or chickens. The average of a year's experience might lead to some satisfactory conclusion.

The *disposition* of the cock and hens should likewise become a subject of careful observation. Some cocks are of an unsocial, *unconjugal* disposition—will persecute and maltreat their hens, and will, if even they leave *them* alone, direct their domineering practices towards the younger inmates of the poultry-yard.

It is often necessary to change the cock, or replace one removed by death, and I must caution my readers to manage this with the utmost possible circumspection. Poultry, although naturally gregarious, are by no means indiscriminate in their attachments, and hens will not, in every instance, admit the company of a new husband when his predecessor has been removed.

Sometimes you will suffer annoyance from the pugnacity of your cocks. This pugnacity is said to arise from an unusually amorous temperament, and a consequent jealousy of disposition. Mascall, or rather his original, Columella, recommends, as a cure for this—"To slake that heat of jealousie, he shall slitte two pieces of thick leather, and put them on his legges, and those will hang over his feete, which will correct the vehement heat of jealousies within him." And M. Parmentier confirms this direction, adding, that "such a bit of leather will cause the most turbulent cock to become as quiet as a man who is bound hand and foot."

Although the cock can by no means boast much of the melody of his voice, he will on no account suffer himself to be *out-crowed* if he can help it; hence, you may observe a cock pause after each crow, in order to ascertain if he be answered by a rival, and the succeeding vocal attempt will, if possible, be yet louder and more discordant.

Cocks and hens are both fond of cleanliness and order in their *plumage*, and are, especially the former, constantly pecking and pruning their feathers. It was formerly, but erroneously, supposed, that during this process an oily fluid, secreted in the gland near the tail, was extracted from its receptacle by the pressure of the beak, and then disseminated over the remainder of the plumage, as a process necessary to render the feathers waterproof. In order to dissipate this illusion, I need only observe, that the *tail-less* fowl, though they are destitute of that part of the body where this gland is situated, and have, consequently, no oil to extract, go through precisely the same process of pecking and pruning, and their feathers are just as much waterproof as those of any other fowl. In my opinion, this fondness of pecking and pruning is partly a provision of nature, designed to relieve some irritation in the skin, and thus conduce to health, and partly proceeds from a pure love of cleanliness and regularity in the plumage, inherent in all varieties of fowl.

In the choice of a hen for sitting, look for a large bird, with large, wide-spreading wings. Though *large*, however, she must not be heavy, nor *leggy*. No one of any judgment would set a Malay hen, as, in such case, not only would many eggs remain uncovered, but many, also, would be trampled upon and broken. Elderly hens will be found more willing to sit than young and giddy pullets; indeed, the latter should never be allowed to sit, until, at least, the second year of their laying.

The Spanish fowl are not generally good sitters; but they are excellent layers. The Dorking reverse the order, being better sitters than layers; and these qualities will also be found to extend pretty generally to hens partaking of the prevailing colors of these two varieties, the black being usually the best layers, and but careless or indifferent sitters; while grey or checkered hens (especially such as have light colored legs) are the best you can procure for sitting hens.

You will be informed of a hen's anxiety to sit, by a peculiar change in her voice to a distinctive *cluck*, which continues after hatching, until the chickens no longer require her maternal care. The heat of the hen's body is also materially increased; hence, when it is desired to check a hen's anxiety to sit, the common practice for allaying this heat is immersion in cold water.

If you entertain doubts of the steadiness of the hen you desire to set, try her constancy by placing her for a few days on some pieces of chalk shaped so as to resemble eggs, or put her on three or four eggs of little or no value.

If you desire to have chickens produced at some particular time, when you have no hen ready to sit, you may induce the desire of incubation by stimulating food—such as toast, or dry bread steeped in good ale, well-boiled oatmeal porridge, with a little Cayenne pepper mixed through it, or hard-boiled eggs, and fresh raw meat, cut small. Fomenting the belly with vinegar, in which pepper has been steeped, is a good practice. But do not suffer any one to persuade you to pluck off the feathers, or to use nettles—practices more cruel than efficacious. Artificial warmth is also never to be lost sight of.

If you find a hen soon tire, or become impatient of sitting, only give her about half the usual quantity of food, and then, when she returns to the nest, feed from the hand with such dainties as you have found to be her favorites. Some will recommend the food to be placed within the hen's reach, in order that hunger, at all events, may not be a means of inducing her to leave her important post. It is not, however, hunger that induces the impatience to which I have alluded; and this total deprivation of exercise is most prejudicial to the poor bird's health. For the first and last week of incubation, however, the hen should only be allowed to quit the nest once daily, and should not be longer than ten minutes absent from the eggs.

Some hens, on the other hand, are as obstinately constant in their sitting as those I have been describing are the reverse; and birds possessing this temperament, will frequently sit until they half starve themselves, if not prevented. Mr. Lawrence says, that he has had hens which, under these circumstances, reduced themselves to such a pitch of weakness as even to *faint;* and, after the chickens were hatched, to be so weak as to be scarcely able to attend them.

Markham scouts the idea of any hen sitting too long, but he is in error. I would not, as some do, recommend such a hen to be fed upon her nest, but I would remove her at proper intervals, and coax her to eat by presenting her with delicacies. If she consent to eat a sufficiency, drinking will be sure to follow. I may here observe, that if a hen acquire the evil habit of breaking and eating her eggs, boil an egg hard, break away a little of the shell, and give it to her while hot. If she peck at it, and, of course, burn herself, you may reckon upon having cured her of her vicious propensity; but should the first painful lesson prove ineffectual, try a second. You will seldom or never have to resort to a third. I think that experience justifies me in arriving at the conclusion that this habit originates in a craving for *calcareous* matter, which I have already stated to be necessary to the well-being of fowl. If your hens be supplied with chalk and sand your eggs will not be touched.

To preserve eggs for hatching, pack them with the small end downward in sand, wood ashes, turf, oats, or other material, for excluding air. But if they are to be kept any length of time, dip them, when new laid, in oil or pure hog's-lard warm—not hot; rub the greasy substance into the pores with the finger, and then pack them with the small end downwards in a box or barrel. For a sea voyage, a coat of varnish would be an experiment worth trying. Care should be taken to push them closely, so that they may be shaken as little as may be.

CHAPTER VI.

SELECTION OF EGGS FOR SETTING—THEIR MANAGEMENT DURING INCUBATION—AND TREATMENT OF THE CHICK AFTER HATCHING.

In selecting eggs for setting, bear in mind what I have said as to the number of hens that the cock should associate with; and choose such eggs as you have reason to believe have been rendered productive. Those of medium size, that is to say, the average size that the hen lays, are most apt to prove prolific. Sketchley tells us that he has always found the round egg to contain the female chick, and that of oblong shape, the male. This, however, though it may have been newly discovered by Sketchley, was known to Columella and Stephanus. If you examine the egg between your eye and a candle, you will be able to discern the position of the vacancy caused by the little air-bag at the blunt end of the shell. If this be in the center, say these authors, the egg will produce a cock; if at one side, a hen. This doctrine, however, has long been abandoned by physiologists, and upon the best authority; nevertheless, though I have no faith in those who pretend to tell the sex of the chickens from the eggs, you may form a very fair judgment if your eggs are impregnated, from their specific gravity. Put them into a bowl of tepid water, and reject such as do not sink to the bottom. Choose, also, such as present a marked disparity of size between the two ends; and while collecting, keep the eggs dry, clean, and in a well ventilated part of the house. Such as are equal in size at both ends, usually contain two yolks; and these, be it observed, instead of producing twin-chickens, as might naturally be expected, commonly produce monstrosities: reject them. The number of eggs to be placed under a hen is from nine to eleven. The number is, however, of course, dependent on the size of both eggs and hen; an *odd number* is to be preferred, as being better adapted for *covering* in the nest. Be sure that they are all fresh; and carefully note down the day on which you place them beneath the hen. Never turn the eggs; the hen can do that better than you. About the twelfth day of incubation, you may be enabled to reject such eggs as are unfruitful. For this purpose, hold the egg between your hands in the sunshine; if the shadow which it forms, waver, keep the egg, as the wavering of the shadow is occasioned by the motion of the chick within; if it remain stationary, throw it away. If your eggs have been recently laid, the chick will be developed earlier than otherwise; if they have been very fresh, you will, about the sixteenth day, if you apply your ear to the egg, hear a *gentle piping noise* within; if the eggs have been stale, this will not be perceptible until about the eighteenth day; and, at this time, the yolk, which had previously lain outside and around the chicken, will be gradually entering into the body of the bird. This serves as nourishment to the little prisoner until his subsequent efforts shall have set him free. From this period let your attention be assiduous, but, at the same time, *cautious;* for the hen has heard this cry before you have, and all her maternal anxieties and tenderness are, from that moment, so greatly augmented, that any unnecessary interference will only tend to irritate her.

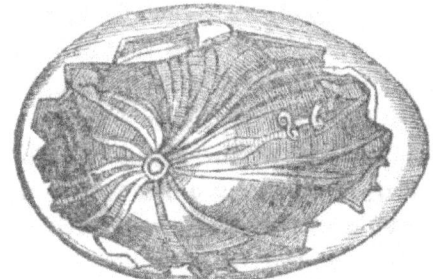

FIRST STAGE IN INCUBATION.

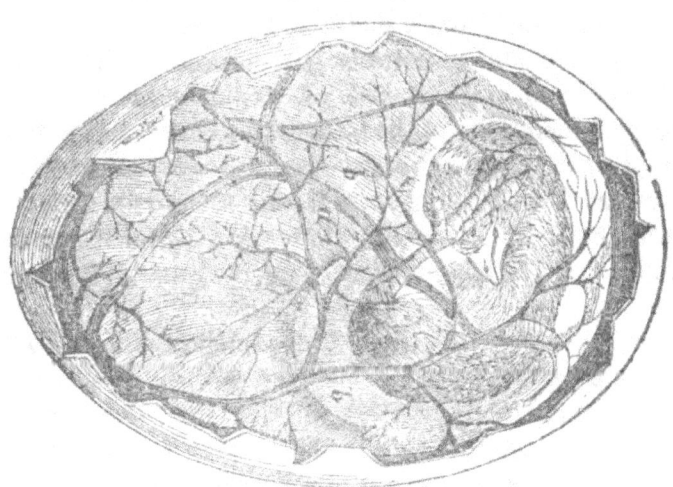

MIDDLE STAGE IN INCUBATION.

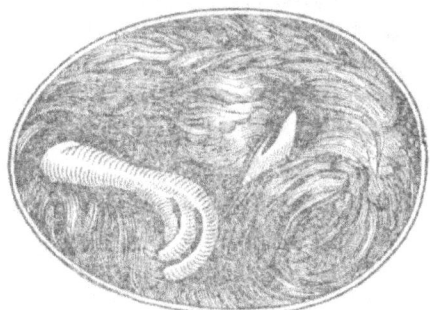

CHICK JUST BEFORE HATCHING.

Eggs during the process of hatching, broken to show the means for supplying nutriment to the chick.

The position which the chick holds within the egg, is apparently anything but advantageous for the work of breaking forth; and, hence, if the youngling be weakly, artificial aid is sometimes necessary. This position would, indeed, almost induce one to regard the liberation of the chick by its own unassisted efforts, as an impossibility. I shall describe it briefly :—The neck slopes toward the belly, to about the centre of which comes the head; the head lies beneath the right wing, just as that of a sleeping bird; the feet are gathered up somewhat like those of a fowl trussed for the spit, and the claws bend backwards, till they almost touch the head; and it is in this confined position that the shelly wall of the prison has to be broken through. It must, therefore, be anything but easy work for the little chick. The process of effecting the breaking of the shell, is a succession of taps from the beak, by which first a crack or *star*, with many cracks diverging from it, takes place; a hole is soon effected, the sides gradually chip away, and the chicken emerges from its new sphere of being. Sometimes the little bird, on proceeding to leave the broken shell, unexpectedly finds itself retained in its place by some accidental or irregular circumstance. The shell may, for instance, have been well cracked, and yet its lining membrane may be so tough as to defy all the efforts of the inmate to rupture it, and thus still present a barrier, and often, without assistance, an insurmountable one. Some chickens *waste their time* striving to tear this membrane before they have made a sufficient crack in the shell. These had better not receive assistance; they will speedily find out their error, and go to work in a proper manner.

In every case *look through the egg* before helping the chick. That chicken which comes out before the *whole of the yoke* has been absorbed, will assuredly prove to be an unhealthy, weakly, little wretch, and will speedily die. A chicken must, previous to leaving the shell, have imbibed such a portion of nutriment as will, at least, serve it for four-and-twenty hours afterward: it is for this that the yolk is designed. Any unusual excess of light, or any injudicious interference with the eggs toward the close of incubation, will nearly always result in causing the chicken to strive to get out too soon, and thus often occasion the loss of numbers.

Neither are all shells, nor all membranes, of an equal thickness, and some are even preternaturally obstinate; hence another difficulty the chick has to experience.

Some poultry keepers will dip the eggs into warm water the day before they think they will be pecked at. This produces no perceptible difference in the consistence of the shell; and I object to the practice, not only on the score of its total inutility, but as being likely to injure the present health of the chick; and the warmth is likewise specially calculated to produce another difficulty connected with its egress, viz., that of being *glued* to the shell, the white of egg—the *albumen* which surrounds the chicken in the shell—being convertible by heat into a kind of GLUE.

The following is, perhaps, the only case in which interference can prove useful:—When you find the fracture on the outside of the shell remaining the same for five or six hours, and when, on examining the edges of this fracture, you find them dry and unmoistened by any fluid, you may conclude that assistance is called for, and may proceed to render it, but, of course, with all possible caution. The best mode to be adopted on such occasions is to imitate, as nearly as possible, the natural efforts of the chicken itself, which may be done by sharp, short strokes with the back of a knife or key; or, what is better than either, the point of a pair of scissors. Be, however, gentle, firm, and deliberate, and take care lest you penetrate the cavity of the egg. Having succeeded in making a sufficient opening in the shell, you may, by a careful and tender use of your fingers, extricate the chick. Sometimes a few scales of albumen, or of the lining membrane of the egg, may remain on the bird's plumage for some days. Do not be uneasy about them. Leave them alone, and as they dry they will fall off themselves. In affording your assistance to the embarrassed chick, be extremely tender with your fingers. You may otherwise often kill when your intention is only to cure. I would be disposed to permit at least *eight hours* to elapse before I resorted to mechanical means of interference. A chick so weak as to perish before that time, is not worth striving to extricate; and, on the score of humanity, its death within the shell will be less painful than after quitting it.

For about twenty-four hours after birth, the chick not only can do well enough without any extraneous nourishment, but will positively be far more likely subsequently to thrive if left alone. The next day they may be fed with crumbs of bread, eggs boiled hard and chopped fine, or cold oatmeal porridge well boiled. After that period, no harm can arise from turning your new brood in among older chicks that already feed themselves. They will then ordinarily follow the example of the rest, and peck away at whatever is going. In the first four days they require food at least hourly, to supply the rapid increase in bulk and feathers. Damp is fatal to them. If the breed is a fine one, however, they will do better with the hen, partaking of the natural food she scrapes together for them.

Although I have mentioned yolks of eggs, boiled hard, and broken down with crumbs of bread, as food for young chickens, I consider this treatment to be needlessly expensive, except in particular cases; and I have found plain crumbs, or cold meal porridge, that has been very well boiled, and not burned, do nearly as well. Small grained meal, given raw, or slightly scalded, and suffered to cool down to a very low degree of tepidity, will also be found useful and good. Do not forget that, in all probability, thirst will be present before hunger, and there ought, therefore, always to be a flat, shallow pan or plate of clean spring water left within reach, and the hen herself, glad of a little refreshment after so long a task, will usually lead the way to it.

If the chickens be hatched during cold weather they will require artificial warmth, or, at the very least, comfortable housing. The kitchen of a farmhouse will afford this in perfection. Recollect that setting your hen in, or at the approach of, winter, is stark folly; freedom from annoyance, comfortable housing, and a sheltered walk, are all that they require—an hour's sunshine is worth more than a year's wrapping up in tow. If your chicks be very weakly you may cram them with crumbs of good white bread steeped in milk; but at the same time recollect that their little crops are not capable of holding more than the bulk of a pea—so rather under than over feed. If your hen have been much exhausted by hatching, you will do well to cram her with crumbs of bread steeped in diluted spirits or ginger cordial.

The following hatching table exhibits the period of incubation with the denizens of the poultry yard:—

	Number of eggs.	Days.
Swan	5 to 10	42
Goose	12 to 15	30
Duck	12 to 15	30
Turkey	15 to 20	31
Peafowl	5 to 7	29
Guinea Fowl	7 to 9	30
Hen	9 to 13	21

CHAPTER VII

VARIETIES OF THE DOMESTIC FOWL.

The varieties of the Domestic Fowl most desirable in an amateur's collection may be classed as follows:—

1. THE MALAY FOWL, from its size and strength, is admirably adapted for crossing with the Dorking and other native breeds.
2. THE JAVA FOWL, nearly resembling, and in the opinion of some, identical with, the Malay.
3. THE COCHIN CHINA breed, equal in most respects, and more prolific than the Malay.
4. THE SPANISH FOWL, perhaps the best breed known for laying.
5. THE POLISH FOWL, a noble and very beautiful bird, and an excellent layer.
6. THE SPANGLED VARIETIES, including the whole class of GOLD and SILVER SPANGLED, known in different countries as Spangled Hamburghs, Every day Dutch, Bolton Bays, Bolton Greys, Chittyprats, Creoles, Corals, &c.
7. THE SPECKLED and WHITE DORKING, the most delicate of all the varieties for the table.
8. THE SUSSEX FOWL, most probably a variety of the Dorking.
9. THE GAME FOWL, graceful of form and plumage, with undying courage, and excellent for crossing with common varieties.
10. THE PHEASANT FOWL, erroneously said to originate in a cross with the Cock Pheasant.
11. THE BANTAMS, more remarkable for their beauty than any other quality.

I. THE MALAY FOWL.

The Malay Fowl has, as its name implies, been brought originally from the peninsula of that name at the southern point of the continent of India. He stands very high on the legs, is long-necked, serpent-headed, and is in

color usually dark brown, streaked with yellow, sometimes, however, white; his form and appearance are grand and striking in the extreme, and he is no small embellishment to the poultry-yard. This fowl is also frequently called the *Chittagong*.

The kind of Malay fowl, however, that were originally imported, were by no means such birds as I could recommend to the notice of the breeder, their size possessing too much offal, as neck, legs, and thighs, and the flesh, moreover, being dark-colored and oily. Another variety has been since introduced, which is well worthy of our attention. As a cross, this Malay has, indeed, proved a most valuable addition to our poultry-yard, the cross-breed possessing all the hardiness of our native domestic fowl, with the gigantic size of the foreign stock.

II. THE JAVA FOWL.

Resembling the Malay in shape, but presenting, in portions of its plumage, the coloring of the Dorking. I hold this, its common appellation, to be a misnomer, and regard it as the result of a cross between a Malay and

Dorking or Spanish. In qualities it resembles the Malay, but s not so val uable as a cross with other breeds.

THE SHAKEBAG.

A good many years ago, there used to be a variety of fowl much in request in England, called the "shakebag," or the "Duke of Leeds' fowl," his grace, of that name, about 60 or 70 years ago, having been a great amateur breeder of them. These fowl were as large as the Malays, but differed from them in the superior whiteness and tenderness of their flesh, as also in their very superior fighting abilities. The name of this fowl seems to have arisen from the old practice of cock-fighting, when *the fancy* used to challenge *all comers* having their cocks concealed in a bag, and the tremendous size and power of the Duke of Leeds' fowl proving so far superior to all competitors, thus usually insuring conquest, and eventually obtaining for it the name, *par excellence*, of *shakebag*, since corrupted into SHACKBAG.

This fine bird was not unfrequently substituted for a turkey, and this to the great convenience of poulterers and inn-keepers.

This "shakebag" or "shackbag" fowl, so lauded by Mowbray, but with the real origin of which he has confessed himself unacquainted, unless, indeed, as an improved breed of dunghill, would appear, if we can judge from the description of Dixon and other writers on poultry, to have been neither more nor less than an offshoot of the great Paduan, Polish, or Jago fowl, the immediate domesticated descendant of the "Gallus giganteus," already described; and I have particularly to request my readers on no account to confound it with the Malay. This fowl would, indeed, seem to have been almost identical with the great wild bird of Sumatra, but it is now altogether unknown to the London dealers. This same fowl was described, about two centuries and a half ago, by Aldrovand, as "very handsome, adorned with five different colors—viz., black, white, green, red, and yellow; body black, tinged with green, tail of the same color; base of the feathers white; some quill feathers of the wings white above; the head adorned with a black crest."

III. THE COCHIN CHINA FOWL.

This gigantic bird has been only very recently introduced into Great Britain and America. The breed have since become comparatively well known and diffused.

This variety of fowl so far surpasses, both in size and power, all that we have ever yet seen in the shape of poultry, as to have led many persons not conversant with zoology, on first viewing them, to refer them to the family of BUSTARDS. They are, however, genuine poultry. Their general color is rich glossy brown, or deep bay; on the breast is sometimes found a marking of a blackish color, and of the shape of a horse-shoe. The horse-shoe mark on the breast is not an infallible sign of the breed. The comb is of a

VARIETIES OF THE DOMESTIC FOWL. 39

THE COCHIN CHINA COCK.

THE COCHIN CHINA HEN.

medium size, sometimes, but not always, serrated—but not deeply so; and the wattles are double. Besides their gigantic size, however, these fowls possess other distinctive characteristics, among which I may mention, as the most striking, that the wing is jointed, so that the posterior half can at pleasure be doubled up and brought forward between the anterior half and the body. The birds can do this at pleasure; and the appearance the manœuvre imparts to their form has procured for them the title of "ostrich fowl." The flesh is white and delicate. The eggs laid by the hen of this variety are

SHANGHAIS.

large, of a light chocolate color, and possess a very delicate flavor. They are very prolific, not unfrequently laying two and occasionally even three eggs on the same day, and within a few moments of each other. The Cochin China fowl is well known in America under the name of Shanghai, being the same fowl with another name.

IV. THE SPANISH FOWL.

This fowl is clad in black plumage, but possesses quite the reverse of black flesh. I regard these birds as the result of the highest possible artificial culture, and adduce, in support of my opinion, their unusually large comb and wattles, characteristics not commonly to be met with among the primitive varieties.

The Spanish fowl is, perhaps, a little inferior in size to the old "shake-bag," but in every other quality, wherein excellence and value are to be looked for, it is more than that bird's equal. The color of the Spanish fowl is a glossy black, and the feathers of the legs, thighs, and belly are particularly decided in their hue, and of a velvety aspect. It is a stately bird, and of a grave and majestic deportment, and is, in either utility or beauty, to be surpassed by none of its congeners. One of the most striking characteristics of this fowl is a white cheek, and the comb and wattles are singularly large, simple, and of a very high color; the feet and legs are of a leaden color, except the soles of the feet, which are of a dirty fleshy hue. A full grown cock will weigh about 7½ lbs.; the hen about 6½.

THE SPANISH FOWL.

This is a fowl well deserving the attention of the breeder, and present no peculiarities of constitution that would suggest difficulties in either hatching or rearing. As table birds they hold a place in the very first rank, their flesh being particularly white, tender and juicy, and the skin possessing that beautifully clear white hue, so essential a requisite for birds designed for the consumption of the gourmand. The hens are likewise layers of the first order; and of all naturalized or indigenous varieties of fowl, with the exception of the Columbian, these lay the largest and the best flavored eggs. They are, besides, prolific, extremely easily fed, and, in short, I know of no fowl I would rather recommend to the notice of the breeder; but let me here observe, that spurious specimens of this fowl are often in the market, which will occasion, perhaps, an equal outlay at their original purchase—will decidedly cost as much to feed—be, perhaps, harder to rear, but will most unquestionably not bring in an equal return in the way of profit. By applying, in the first instance, to a breeder of known respectability, you will avoid much disappointment; and though you may conceive the price demanded of you to be high, it may not, perhaps, at the same time, be higher than what you might have foolishly paid for a bad article; and even should you have to pay an extra price, do so willingly, and, recollecting the old proverb, avoid being "penny wise and pound foolish."

THE COLUMBIAN.

A very noble fowl, presenting the appearance of a cross between Spanish and Malay, but possessing so much nobility and stateliness of aspect that I am loth to regard it otherwise than as a distinct and very primitive variety. The eggs are particularly large. My fowl, of this breed, lay eggs averaging in weight from 4¼ oz. to 4½ oz., seldom, however, laying more frequently than every second day. These fowl are natives of Columbia, on the Spanish main in South America; and I think it not improbable that they are the origin of the breed now known as "Spanish."

V. THE POLISH FOWL.

The *Golden Spangled* is one of no ordinary beauty; it is well and very neatly made; has a good body, and no very great offal. On the crest, immediately above the beak, are two small fleshy horns, resembling, to some extent, an abortive comb. Above this crest, and occupying the place of a comb, is a very large brown or yellow tuft, the feathers composing it darkening towards their extremities. Under the insertion of the lower mandible, or that portion of the neck corresponding to the chin in man, is a full, dark-colored tuft, somewhat resembling a beard. The wattles are very small. In the golden variety, the hackles on the neck are of a brilliant orange, or golden yellow; and the general ground-color of the body is of the same hue, but somewhat darker. The thighs are of a dark brown, or blackish shade, and the legs and feet are of a bluish gray. The full grown cock weighs about six pounds, and the hen five and a-half pounds; the eggs moderate in size, and very abundant.

In the *Silver Spangled* variety, the only perceptible difference is that the ground-color is a silvery white. The extremity, and a portion of the extreme margin of each feather, are black, presenting, when in a state of rest, the appearance of regular semicircular marks or spangles; and hence the name of "Spangled," the varieties being termed gold or silver, according to the prevailing color being bright yellow, or silvery white. In mere excellence of flesh, and as layers, they are inferior to the Dorking or Spanish varieties.

Of the Polish fowl there are several subvarieties. The Polish fowl is, perhaps, the most unchanged from the primitive stock of any we are now acquainted with, being beyond doubt the immediate and almost unmixed decendant of the "Gallus giganteus," or great wild cock of Sumatra. The varieties of Polish fowl are—

GOLD SPANGLED POLISH HEN.

I.—*The Spangled Polish.*—A bird of extraordinary beauty, extremely scarce, and very difficult to be procured. This fowl presents a symmetrical and regular combination of the following colors, viz. :—A bright orange, a clear white, a brilliant green, and a jetty black, softened down with a rich and pure brown, every feather being tipped with white, so as to produce the effect whence has been derived the term of SPANGLED. The color of the hen is a prevailing golden yellow, with white spangles, like the cock. In the cock the thighs are black, and are, likewise, though in a less degree, marked and spangled with black and golden yellow. The hinder end of the body is furnished with green and orange-brown hackles, and the tail is carried well up. The flesh of these birds is of good quality, and they are very prolific. They also fatten quickly, and have, by some, been compared to the Dorking for similarity of flesh and other excellences of quality.

WHITE-CRESTED BLACK POLANDS.

II.—The second variety of the Polish fowl is the well-known black fowl, with a white tuft on the crown. These birds were brought from St. Jago by the Spaniards, to whom they owe their first introduction into Europe. Their color is a shining black, and both cock and hen have the white top-knot. The head is flat, surmounted by a fleshy protuberance, out of which spring the crown feathers constituting the tuft. These are remarkably good layers, and will, if kept warm, lay nearly throughout the year; and it is this cause, probably, that has induced Mowbray and other writers to confound them with the Dutch breed, which, from a similar circumstance, have been styled "Every-day layers."

III.—This variety of Polish fowl is the most pure and unmixed of the three: it is, indeed, to all appearances, the uncontaminated descendant of the great fowl of St. Jago. Its color is a brilliant white, with a jet black topknot. This variety was described by Aldrovande, and more recently by Dr. Bechstein. I have never myself seen a specimen of the breed, and have every reason to suppose it to be extinct, or very nearly so. Applications have been made to several persons in both Germany and Poland, connected with the poultry fancy, for the purpose of procuring specimens of these birds at any cost, but the answers returned were, without one exception, that they were no longer to be had.

VI. SPANGLED VARIETIES.

Gold and Silver Spangled Hamburgs.
Dutch Penciled.
Dutch Every-day Layers.
Bolton Grays.
Gold and Silver Dutch.
Gold Bolton Bays.
Chittiprats.
Creoles.
Prince Albert Fowl.

Much confusion seems to exist with regard to the spangled varieties of the Domestic Fowl. The truth seems to be that the spangled fowl have been introduced without much attention being paid to their origin, and breeders have given them the names they thought most descriptive of their appearance and qualities—and thus run hastily through the description of the spangled varieties, as given by Mr. Dixon and other writers;—in the south of England a variety exists called the *Coral*, or *Creoles*, to which the Penciled Dutch of Dixon is the nearest approach.

In the neighborhood of Keighley, in Yorkshire, and on the borders of Lancashire, the Bolton Greys are called "Chittiprats," or "Cheteprats," and prizes given to them as handsome, hardy, and excellent layers. In other parts of the kingdom they are known by the name of "Moonies." The so-called Prince Albert's breed are Bolton Greys, said to be crossed with game blood, and not easily to be distinguished from the Silver Spangled Hamburgh. Bolton Bay is another provincial term for the Golden Hamburgh, as Bolton Grey is for the Silver.

It is obvious, from these confused statements, that the various spangled races of Domestic Fowl have been so intermingled as to render it next to impossible to discriminate between them.

DUTCH EVERY-DAY LAYERS.—Frequently confounded with the preceding. Instead of being destitute of comb, and carrying in its place a tuft of feathers on the crown, the cock of this interesting variety possesses what is called a rose comb; that is to say, a comb formed of a great number of folds of single comb, united into one broad, serrated, and fleshy mass. The color of the cock is, as usually occurs, more brilliant than that of the hen. His body is of a fine, reddish-brown hue, with neck hackles of a bright and rather deep golden yellow. These birds present, likewise, two distinctly-marked varieties, the difference, however, depending chiefly on color. When, as I have described, the color of the body is a golden yellow, streaked or spangled with blackish, or deep brown markings—an appearance caused by the dark color of the ends of the feathers—the bird is styled the "Golden Spangled;" and when the ground color is white (the other circumstances of shading remaining the same), the bird is styled the "Silver Spangled."

These fowl have received the name of "Every-day" or "Everlasting Layers," from the circumstance of their unwillingness to hatch, in consequence of which they lay an egg nearly all the year through, and, if properly cared for, and warmly housed, even amid the frost and snow of the most inclement winter. Some say that the eggs of these fowls are not in general so large as those of ordinary poultry, nor equally substantial and nutritious. This might, indeed, considered theoretically, seem a very obvious consequence of so unsound a demand upon the bird's natural resource; but I think that there is really no such remarkable difference.

THE BOLTON GREYS.—In general form they resemble the Dorking, except that they are longer in the body; the color elegantly penciled in black. A variety called Bolton Bays, from that color, have precisely similar pencilings upon the bay color. Mowbray, quoting the Rev. Mr. Ashworth, vicar of Tamworth, says of the Bolton Greys :—"They are small in size, short in the leg, and plump in the make; the color of the genuine kind invariably pure white in the whole lappel of the neck; the body white, thickly spotted with bright black, sometimes running into grizzle, with one or more black bars at the extremity of the tail. They are chiefly esteemed as very constant layers, though their color would also mark them for good table fowl." Mowbray also calls them Corals—why, does not appear, unless they are synonymous with the Creole of other parts of the country. In Yorkshire, the same birds are called Chittiprats.

THE BARBARY FOWL.—Now naturalized in Spain; the specimen that I describe was brought recently from that country. It is very tall, remarkably heavy, with not much offal, and a firm, muscular quality of flesh. The comb presents a most singular appearance—viz., that of two large and fleshy combs growing up together, and enclosing a smaller and apparently abortive comb between their folds. The color is a prevailing black, with some green and brown markings upon the wings; it is booted and feathered upon the legs, like the Bantam, and thus clothed to the very toes; the cheek or ear-piece is white, like the Spanish breed. It is a bird of vast body, and almost gigantic proportions, displaying great boldness of carriage and confidence of demeanor.

VII. THE DORKING FOWL.

The Dorking would appear to owe its name to its having been chiefly bred in a town of Surry, of the same appellation. That the peculiarity of five toes, or, in other words, of two hind toes instead of one, is to be regarded as a distinctive character of the breed, is by some writers questioned, and by others wholly denied. For my part, I should say, that whenever this characteristic is absent, a cross has been at work.

WHITE DORKINGS.

I do not, however, mean to assert that this possession of two hind toes instead of one, has never occurred in any other family of fowl except those bred at Dorking, in Surry, for Aristotle has mentioned the existence of a similar peculiarity among certain fowl in Greece, and both Columella and Pliny assert the existence of such in their time in Italy, so also does Aldrovand; and these authors lived hundreds of years ago; and, oddly enough, these breeds were remarkable, as are our own Dorking, for being good layers and good sitters.

The color of the Dorking is usually pure white, or spotted or spangled with black; these colors sometimes merge into a grey or grizzle. The hens weigh from seven to nine pounds; stand low on their legs; are round, plump, and

VARIETIES OF THE DOMESTIC FOWL. 47

shor' in the body; wide on the breast, with abundance of white juicy flesh. The hens are generally good layers, and their eggs, though smaller than the egg of the Spanish and Polish breeds, are of good size and well

COLORED DORKINGS.

flavored. These birds have been long prized, and it is now many years since their superiority over our ordinary domestic varieties was originally discovered and appreciated; they were first noticed, and the variety adopted, by the Cumberland breeders, whence they were soon brought into Lancashire and Westmoreland, and gradually spread over all England. Whether, however, from injudicious treatment, or imperfect feeding, or change of climate, or from whatever cause, it is certain that, when met with far from their native place, they appear greatly to have degenerated from their original superiority of character. In this, and all other varieties of fowl, fresh blood should be introduced from time to time, or the breed degenerates.

VIII. THE SUSSEX.

This is but an improved variety of Dorking, similar in shape and general character, usually of a brown color, but possessing the advantage of wanting the fifth toe: I say advantage, for the Dorking fowl frequently becomes diseased in the feet, the cocks especially, in consequence of breaking the supplementary toe in fighting.

IX. THE GAME FOWL.

The Game fowl is one of the most gracefully-formed, and most beautifully-colored of our domestic breeds of poultry; in its form and aspect, and in the extraordinary courage which characterizes its natural disposition, it

exhibits all that either the naturalist or the sportsman recognizes as the *beau ideal* of high blood; embodying, in short, all the most indubitable characteristics of gallinaceous aristocracy.

We do not possess any very satisfactory record of the original country of the Game fowl; but I am disposed to cede that honor to India, the natives of which country have always been remarkable for their love of cock-fighting; and we also know that there still exists in India an original variety of game cock, very similar to our own, but inferior in point of size. As to the date or occasion of their first introduction into the British islands, we know nothing certain; but it is probable that we owe it to the invasion of Julius Cæsar, the Romans having been very fond of the sport of cock-fighting.

VARIETIES OF THE DOMESTIC FOWL.

The Game fowl is somewhat inferior in size to other breeds, and in his shape he approximates more closely to the elegance and lightness of form usually characteristic of a pure and uncontaminated race. Amongst poultry he is what the Arabian is amongst horses, the high-bred short-horn amongst cattle, and the fleet greyhound amongst the canine race.

The flesh of the Game fowl is beautifully white, as well as tender and delicate. The hens are excellent layers, and although the eggs are somewhat under the average size, they are not to be surpassed in excellence of flavor. Such being the character of this variety of fowl, it would doubtless be much more extensively cultivated than it is, were it not for the difficulty attending the rearing of the young, their pugnacity being such, that a brood is scarcely feathered before at least one-half is killed or blinded by fighting.

The beauty of form and brilliancy of color displayed in the Game fowl, renders the breed very desirable; they are of all colors, and each variety seems to have had its patrons, the rule being to mate the cock with hens of the same feather. The brood cock for purposes of battle, says this authority, "should have every feature of health; such as a ruddy complexion, feathers close and short, flesh firm and compact, breast full, yet taper, and thin behind, full in the girth, well coupled, lofty and spiring, a good well-developed thigh, the beam of the leg strong, a large quick eye, beak strong and crooked."

GAME-COCK.

"In the choice of your hens," says Spetchly, "let them be rightly plumed to the cock; nor let your choice fall upon those that are large, but rather suffer the cock to make up for the deficiency of the hen in size. In shape they should be similar to the cock, lofty necks, short and close feathered. A true blood hen is clean and sinewy in the leg, the body compact and well proportioned, a well-set thigh, with long, clean, and taper toes." Having selected a cock, place with him from four to six hens, bringing them together in November or December. If he is young, the hens may be full-grown—if a two year's old, then the hens may be young pullets, supposing a strong and vigorous breed is desired. Have, however, a marked attention how he bears himself to all his hens, as it frequently happens that one or other of them falls under his displeasure, in which case she should be removed.

In selecting eggs for setting avoid the earliest ones, as well as the last; choose the best shaped eggs and mark them to avoid mistakes, and place them under an old game hen if you can procure one, the old being excellent mothers. Their place for sitting should be private, and free from all annoyance or intrusion.

When hatched, the young should be regularly fed, and often, after the first day or two, but in small quantities. Let their food be:—Macerated eggs, boiled hard; crumbs of white bread; lettuce leaves and meadow ants; maggots from grains; steeped oats and small wheat; curds, with new milk; bread, toasted, and steeped in chamber-lye.

The variety of Game fowl are very numerous, and to the uninitiated their designations very unintelligible. For the purposes of combat, a sport now rarely followed by amateurs, the black-reds have been the favored variety. The recognized breeds are, according to Spetchly:—

1. Black Reds.
2. Silver black breasted ducks.
3. Birchen ducks.
4. Dark greys.
5. Mealy greys.
6. Blacks.
7. Spangles.
8. Furnaces.
9. Pole cats.
10. Cuckoos.
11. Gingers.
12. Red duns.
13. Duns.
14. Smoky duns.

"In all these," says Spetchly, "good birds may be found; from them, however, have been raised crosses innumerable, and it is the aim of the fine breeders of the present day to have their birds as much as possible uniform in feather, blood, and constitution. Piles," he says, "have originated from a variety of crosses, which have constituted many of the shades of color; they are not," he adds, "of my selects."

Buffon, and other continental writers, have given this fowl the not unappropriate title of the "English Fowl;" and truly it is in England that the very best specimens of the breed are to be met with.

A correspondent well acquainted with rearing and breeding of game fowl, says, "Four or five hens are quite sufficient to keep company with one game cock; perhaps, it is right to observe, that as hens lay at various seasons of the year, there never should be at any one particular season more than eleven or thirteen eggs collected for hatching. When this is done the chickens will prove to be more spirited and resolute. The month of March is the best month to bring forth game chickens. It is generally understood that when hatched in that month they prove to be the most hardy and constitutional birds. In putting game hens with a cock for breeding, great care should be taken to match the feather as near as possible. You may breed from a cock until he is four years old—that is, if not previously cut up by fighting a battle. One battle, or even two, if easily won, will not injure a cock for breeding; some say it will, but I think not. Pullets should at all times be put to aged cocks, and *vice versâ*, stags to aged hens. The greatest of care should be taken in gathering the eggs, that those of each hen be kept separate, and hatched accordingly.

We should state, in conclusion, that however interesting for their beauty and high courage, game fowl will be very troublesome in a poultry yard of various breed, especially if any other cock is kept; for although their smaller size might lead to the supposition that they would not be the aggressors, this would be a mistake; their indomitable spirit leads them to quarrel with every other bird, and their activity and muscular strength render them dangerous to the largest adversary.

X. PHEASANT FOWL.

Much has been written upon this bird and its origin, and a candid consideration of the entire subject, leads to the conclusion that this is another case of intermingling of different varieties. Certain it is, that no established instance exists, where a cross between the pheasant of the woods and the domestic fowl have ever reached a second generation.

Mr. Whittaker of Beckington, Somerset, describes a breed of what he calls PHEASANT MALAY, which he has kept for seven years. The cock he describes as a large sized bird, of a dark red color, with a small comb; but the beauty of the breed is with the hens, which are of a pheasant color in all parts of the body, with a velvety black neck, the shape of both cock and hen being very good; the neck in both, long and high crested; the legs, and also the skin, is white. The hens have scarcely any comb; the cocks have one extending only a little way backwards. The chickens of this breed hatched in June, succeed better than when hatched earlier; that they are small at first, and being scantily supplied with down, have a naked appearance, and are very susceptible of cold, circumstances which lead him to suspect them to be a recent introduction, and from some warmer climate.

XI. THE BANTAMS.

The original of the Bantam is the Bankiva fowl, a native of Java, several specimens of which are kept by the Queen of England. These are very beautiful, of a perfectly white color, and exceedingly small size, and they exhibit some peculiar traits of habit and disposition that we cannot overlook. Amongst other strange propensities, the cocks are so fond of sucking the eggs laid by the hen, that they will often drive her from the nest in order to obtain them—nay, they have even been known to attack her, tear open the ovarium, and devour its shell-less contents.

As might be inferred, when such a propensity to devour the eggs exists

DOMESTIC FOWL.

WHITE BANTAMS.

in the male bird, the female is a secret layer. In this respect these fowl show their identity with the original bird of Java—the Bankiva cock These birds are both good layers and good sitters.

The fowl commonly known as the Bantam, is a small, elegantly-formed, and handsomely-tinted variety, evidently not remotely allied to the game breed. This bird is furnished with feathers to the toes. There is another variety ordinarily known as Sir John Sebright's fowl, which has its legs

ORDINARY BANTAM.   SEBRIGHT BANTAMS.

perfectly naked to the toes, and approaches in form more nearly to the game breed. The high-bred cock of this breed should have a rose comb, full hackles, a well-feathered and well-carried tail, a stately, courageous demeanor, and should not be quite a pound weight. The favorite color is a golden yellow, the feathers edged with black, the wings barred with purple, tail feathers and breast black. The Bantam possesses high courage, and will fight with great resolution. The attitude of the cock is singularly proud and haughty; his head thrown back so as to nearly touch the upper feathers of his tail. Pure birds of this blood are very rare.

THE CREEPER is also a very small variety of "Bantam," with short legs.

XII. THE TURKISH FOWL

Is another variety of "Bantam," having a whitish body, with black belly and wings, the body streaked with gold and silver, and the legs bluish. The hen is, as usual, of a less showy plumage, her color being white, speckled here and there with black, the neck yellowish, and tail of one color.

XIII. THE JUMPER.

In addition to these diminutive races, there is another mentioned by Buffon, as being so short-legged that they are compelled to progress by *jumps*. These are, however, somewhat larger than the common Bantam, and approach more nearly in size to the Dunghill. They are prolific, as well as excellent sitters, the hen having been known to hatch two broods in succession, without even an intermediate day of rest. These dwarf fowl were described by Aldrovand more than two hundred years ago, and also, much farther back, by the celebrated Roman naturalist, Pliny, under the designation of the *Adrian breed*.

XIV. THE RUMPKIN OR TAIL-LESS FOWL.

This bird is distinguished by the total absence of the caudal extremity. Some suppose it to be a distinct species descended from the wild breed of Ceylon. Among the wild birds the comb is not indented; it is so with the tame; and is, in the latter case, frequently double. Buffon supposed this fowl to be a native of America, but Dixon declares him to have been in error, having been misled by the circumstance of these birds being domesticated very commonly in Virginia. Others have supposed this fowl to be a native of Persia, and Latham even names it the "Persian Cock." It is, however, of very little practical importance whence the rumpkin originally came, the bird possessing neither good flesh nor affording good eggs.

XV. THE SILKY FOWL.

This fowl, remarkable for the silky texture of its plumage, is a native of China, but is likewise to be found in Japan: it is nearly always of a white or cream color. Some modern writers have sought to establish for the silky fowl a claim to be considered a distinct species; but their opinion is evidently erroneous. These fowl are good layers, but the eggs are small. For any practical purpose they are quite useless, and are also carefully to be excluded from the poultry yard, on account of the rapidity with which a cross from them lowers the value of our common poultry, darkening the color of the skin, and causing our birds to deteriorate both in appearance and utility.

XVI. THE SIBERIAN OR RUSSIAN FOWL,

Called by some the Russian, and said to be a native of that country, is distinguished by tufts of dark-colored feathers springing from each jaw, others, longer and fuller, springing from the lower mandible, in the form of

RUSSIAN FOWLS.

a beard. The color varies; some are white, some blue or black, and others are colored like the game fowl. The flesh of this variety is white and good. They are, likewise, good layers, are hardy, and easily fed. This fowl is sometimes colored like the Spangled Hamburgh—some gold and some silver spangled. When thus colored, they are deemed valuable.

XVII. THE FRIZZLED FOWL

Is so called from the crisped and frizzled appearance of its feathers, and not, as some have erroneously asserted, from a corruption of Friesland, at one time improperly conceived to be its native country. It is a native of Java, and other parts of Eastern Asia: it is smaller than our common fowl, is very susceptible of cold, and is, on that account, very difficult to rear. These fowl are particularly sensible of wet, the chickens especially; they are very shy and wild, and, like the Rumpkin, are objects for the attention of the showman rather than of the poultry breeder.

XVIII. THE DUTCH FOWL

Is of a white or grey color, streaked or spangled with black, and excellent fowl, whether as layers or for the table; originally imported from Holland. This is called by Dixon the "Pencilled Dutch Fowl," from its marking. It is not the same as the birds I have already described under the name of "Every-day layers."

XIX. THE NEGRO FOWL

Is a native of Africa, but by no means to be confounded with the "Barbary fowl." The Negro fowl is distinguished by having black comb, wattles, skin, bones, and feathers. The flesh is, however, white and tender. This bird is another good specimen for the curious, but anything but a desirable inmate of the poultry-yard, as, besides being ugly and unprofitable, he has the objectionable quality of speedily causing deterioration among poultry.

XX. THE BARN-DOOR FOWL.

I describe these fowl separately; for, although the designation of "Barn door fowl" may be applicable also to the Dunghill, I regard the former appellation as possessing a far more extended signification.

The Barn-door fowl embrace, of course, several sub-varieties. Few of our high-priced breeds, except in some places the Dorking and the Polish, have, as yet, become so common as to be included in the list; but crosses of the common Dunghill bird with the Malay, Dorking, Polish, or Spanish, are very frequently to be met with.

Dr. Bechstein enumerates eight distinct varieties of barn-door fowl, viz:—

1. The fowl with the small comb.
2. The crowned fowl.
3. The silver-colored fowl.
4. The slate-blue fowl.
5. The chamois-colored fowl.
6. The ermine-like fowl.
7. The widow; with tear-like spots on a dark ground.
8. The fire and stone-colored fowl.

The distinction will be perceived to consist almost solely in color; but the Doctor has omitted another and very ordinary inmate of the farm-yard—viz., the *booted* fowl, represented by the bantam. It will then be seen that the Barn-door fowl, whatever marks of being an original variety it may have formerly exhibited, is now likely soon to lose all such marks from the effect of crossing.

XXI. THE DUNGHILL FOWL.

The Dunghill fowl occupies in the poultry-yard precisely the position of the cur dog in the kennel, being, in fact, the produce of a miscellaneous intermixture of most of the ordinary domestic varieties, and constantly differing in its appearance with the accidents which may have influenced its parentage.

CHAPTER VIII.

THE TURKEY.

THE WILD ORIGINAL.

Linnæus and others have given the turkey the erroneous appellation of "Maleagris gallipavo," under the strange impression that this bird and the Maleagris of the ancients are identical—a very strange error indeed, inasmuch as the descriptions of the Maleagris, given by Athenæus and other classic writers, refer with the most minute accuracy, to the *Guinea fowl;* and in scarcely any single particular can be traced a resemblance to the turkey. The mistake was first observed and pointed out by the French academicians, and is now universally admitted.

Various opinions have been promulgated relative to the original country of the turkey, but it is now ascertained beyond a doubt to have been America; and it is in that country alone that the true original of our domestic turkey is yet to be met with in all its primitive wildness, clothed in its natural plumage, genuinely wild in all its habits, the unreclaimed denizen of the wilderness. As to the medium through which this bird was first introduced into Europe much doubt still exists, and we have, indeed, no authentic proof as to either the period of time, or by what agency that event took place; it is, however, not unreasonable to suppose that the Spaniards, after their discovery of Mexico, where the turkey is known to be indigenous, brought specimens away with them on their return to their own country; and Oviedo, the earliest describer of this bird, speaks of it having been domesticated by the Christian inhabitants of New Spain and the Spanish Main. This proves that the turkey was domesticated by the Spaniards before the year 1526, for in that year was Oviedo's "Natural History of the Indies" published at Toledo. The discovery of Mexico took place in 1518: and when Hernandez shortly afterward described the natural productions of that country, he enumerated amongst them the turkey, distinguishing also the *wild* from the *tame.* In 1530, the turkey was introduced into England; but it seems more probable that we owe its introduction to Cabot's having brought it direct from America, than that we obtained it from Spain; for if the latter were the case, I think it likely that some record of its transmission would remain.

In 1541, we find turkeys enumerated amongst the delicacies of the table, and classed with the crane and swan; but the bird was too important an addition to our stock of domestic poultry to remain very long a rarity. Attention was drawn towards it,—it was bred extensively; and in 1573, we find it mentioned in "Five Hundred Points of Good Husbandry" as forming the staple of the farmer's ordinary Christmas dinner.

The origin of the popular name "Turkey" appears to be the confusion that at first so unaccountably subsisted relative to the identity of the bird with the Guinea fowl, which is really a native of that country, and which was introduced into England from the Levant, and at the time of the introduction of the turkey was still scarce. Some say it arose from the proud and Turkish strut of the cock. An old writer on agriculture, named Googe, (A.D. 1641,) asserts that the turkey and Guinea fowl were unknown in Britain in 1530; but he evidently suffered himself to be misled by a German author, Heresbach, whose treatise seems to have been the basis of Googe's work. Hakluyt (A.D. 1582) mentions their having been introduced "about fifty years back." In 1555, two turkeys and four turkey poults formed part of the inauguration dinner of the serjeants-at-law in London : they cost only four shillings each, while the swans were rated at ten shillings, and capons at half a crown : turkeys could not, therefore, have been very scarce at that time.—*Dugdale, Orig. Jud.* Thus, the turkey would appear to have been introduced into England about the year 1530, and we may conclude that it was brought into France about the same period; for, in "Champier's Treatise on Diet," published in 1560, the turkey is described, and the work is said to have been written upwards of thirty years prior to its publication. In this book, also, the bird is said to have been brought from the "newly discovered Indian islands;" and my readers are well aware that the newly discovered continent of America was at first conjectured to be a portion of India, or an island belonging to it. In 1556, twelve turkeys formed the present offered to the King of France by the burgesses of Amiens. Heresbach states that they were introduced into Germany about 1530, and a sumptuary law made at Venice, in 1557, indicates the rank of those at whose tables they were permitted to be eaten. The turkey was then early appreciated, and his value duly estimated; yet strange to say, not a record remains to lead us to a knowledge of the person to whom the natives of Europe are indebted for so very important a benefit. The turkey has long enjoyed the reputation it now holds, and has been deemed worthy of a place at the most luxurious festivals.

No one who has seen only the domesticated inhabitant of the poultry-yard can form any idea of its wild original. The cock measures about three feet and a half, or nearly four feet, in length, and almost six in expanse of the wings. The skin of the head is of a bluish color, as is also the upper part of the neck, and is marked with numerous reddish, warty elevations with a few black hairs scattered here and there. On the under part of the neck the skin hangs down loosely, and forms a sort of wattle; and from the point where the bill commences and the forehead terminates, arises a fleshy protuberance, with a small tuft of hair at the extremity, which becomes greatly elongated when the bird is excited; and at the lower part of the neck is a tuft of black hair, eight or nine inches in length.

The feathers are, at the base, of a light dusky tinge, succeeded by a brilliant metallic band, which changes, according to the point whence the light falls upon it, to bronze, copper, violet, or purple; and the tip is formed by a narrow, black, velvety band. This last marking is absent from the neck and breast. The color of the tail is brown, mottled with black, and crossed with numerous lines of the latter color. Near the tip is a broad black band, then a short mottled portion, and then a broad band of dingy yellow. The wings are white, banded closely with black, and shaded with brownish yellow, which deepens in tint towards the back. The head is very small in proportion to the size of the body; the legs and feet are strongly made, and furnished with blunt spurs about an inch long, and of a dusky reddish color; the bill is reddish, and horn-colored at the tip.

The hen is less in size than the cock; her legs are destitute of spurs; her neck and head are less naked, being furnished with short, dirty, gray feathers: the feathers on the back of the neck have brownish tips, producing, on that part, a brown, longitudinal band. She also frequently, but not invariably, wants the tuft of feathers on the breast. Her prevailing color is a dusky grey, each feather having a metallic band, less brilliant than that of the cock; then a blackish band and a greyish fringe. Her whole color is, as usual among birds, duller than that of the cock; the wing feathers display less white, and have no bands: the tail is similarly colored to that of the cock. When young, the sexes are so much alike, that it is not easy to discern the difference between them; and the cock acquires his beauty only by degrees, his plumage not arriving at perfection until the fourth or fifth year.

WILD TURKEY.

The wild turkey was formerly found in Canada, and in several districts of the United States, but has been gradually driven backwards as population increased. It is now chiefly to be found in the wilder regions of Virginia, Kentucky, Ohio, Illinois, and Indiana. The wild turkey is, to a certain extent, migratory in its habits; and about the latter end of autumn large flocks assemble, and gradually desert their barren wilds for richer plains. The cocks associate in parties by themselves, and seek for food apart from the hens. The latter remain with the poults, which they take care to keep away from the cock, who is very apt to attack and destroy them.

Flocks leaving the same district all move forward in the same direction. They very seldom take wing unless to escape an enemy, or to cross a river, which latter feat they do not perform without great deliberation, and a great deal of noisy "gabbling." The old and strong birds will fly in safety across a river upwards of a mile in breadth; the young and weakly often fall in, unequal to the effort; but nevertheless usually manage to attain the shore by swimming. On reaching the opposite bank, the flock will generally strut about for a length of time, as if bewildered, and may, during this interval, be readily taken. On arriving at the desired district, they disperse in smaller flocks, composed indiscriminately of cocks, hens, and poults. Their food consists of beech-mast, maize, a fruit called the peccan nut, and acorns. They will also devour such beetles, grasshoppers, young frogs, small lizards, &c., as fall in their way. This is about the month of November, at which season they often incautiously venture too near farm-yards and barns, where great numbers are killed, and form a valuable article of traffic to the settler

Early in March the hens separate again from the herd, roost apart, and carefully shun the cock. They still, however, remain near him; and when a hen utters her call, every cock within hearing responds with his "gobble," "gobble," "gobble." This noisy wooing generally continues for about an hour before sunrise, after which the birds silently alight from their perches, and the cocks strut about with expanded tails, seeking to obtain the favor of their desired mates. They sometimes, while thus employed, encounter each other, in which case desperate conflicts take place, terminated only by the death or flight of the vanquished.

After pairing, the birds remain together for the season, until laying begins, when the hen is again compelled to seclude herself, as the cock would otherwise destroy the eggs. About the middle of April the hen forms her nest of a few dry leaves, on the ground, in some sheltered spot, where it will be concealed from every hostile eye; here she deposits her eggs, to the number of from ten to twenty. They resemble, in size and color, those of the domestic bird. Whenever she leaves the nest, she covers it up with leaves, so as to secure it from observation. She is a very close sitter, and will, also, when she has chosen a spot, seldom leave it on account of its being discovered by a human intruder. Should she find one of her eggs, however, sucked by a snake, or other enemy, she abandons the nest for ever. When the eggs are near hatching, the hen will not forsake her nest while life remains.

The young are very sensible to the effects of damp; hence, after a rainy season, wild turkeys are always scarce. The flesh of the wild turkey is very superior to that of the domestic bird; yet that of such of the latter as have been suffered to roam at large in the woods and plains is in no respect improved by this partially wild mode of life. The wild bird is frequently domesticated in America; but I understand that these individuals are not very steady, and will, on the first opportunity, return to their native haunts.

C. Lucien Bonaparte relates that a gentleman in West Chester County, New York, once procured a young female wild turkey, in order to try the experiment of crossing the breed with the domestic bird; but owing to some accident it did not succeed, and in the ensuing spring the hen disappeared. She returned, however, in the autumn, followed by a large brood, and remained on the farm till the following spring, when she again disappeared, but returned in autumn with a second brood; and this she continued to do for several years.

When the eggs of the wild turkey are hatched under a tame hen, the poults preserve the wild manners of their race, and roost apart from the rest. These are often used as decoy birds, for the purpose of securing the wild ones. The wild turkey is found to thrive better, and fatten sooner, on a given quantity of food, than the tame; and it is well known that the cross between the two is a greatly improved breed as to flesh and capability of taking fat. Some writers have greatly exaggerated the weight of the wild turkey; and some have even asserted that they have met with individuals of sixty pounds weight. M. Bonaparte states the average weight of the hen to be from eight to nine pounds, and that of the cock from fifteen to twenty. A knowledge of the natural habits of the bird is of the greatest importance in guiding us as to its treatment in a state of domestication; and we, accordingly, should avoid condemning to the confinement of close, and often filthy hen-houses, a bird which, in a state of nature, always perches in the open air. Open sheds and high perches are what they require; and their dislike to the mode of housing I speak of may be recognized in the eagerness with which they rush out the instant the door is opened in the morning. The domestic turkey has been known to go wild and remain so for two or more years; and there is no doubt that it would be possible to naturalize them like the pheasant.

Domestication has, in the case of the Turkey, as in that of most reclaimed animals, produced a diversity of color, which by cultivation, whether owing to fancy or some supposed inherent excellence residing in the various tints, has now furnished us with several so-called varieties or breeds, still however, with one exception (the Norfolk), only differing in the prevailing hue of their plumage: thus we have the black, the white, the copper color, the brown, the bronze, and the dusky-grey. They are however, of course, all the descendants of their great American original, of which but one really exists, although F. Cuvier has described (1820) a second species found at Honduras. There is a question whether this actually be a second and distinct species, however, or merely a variety of the wild bird, owing its diversity of aspect to circumstances dependent on locality, and consequent change of habit, combined with difference of climate and other important causes, which we know, in the case of other animals, produce such remarkable effects.

THE TURKEY.

THE DOMESTIC TURKEY.

As to the relative value of the ordinary varieties, it would be almost difficult to offer an opinion; but those who suppose the white turkey to be " the most robust and most easily fattened" are decidedly mistaken, both in theory, as far as analogy may guide us, and in practice, where the certain test of experience has shown to the contrary. The bronze and copper-colored varieties are generally undersized, and are amongst the most difficult of all to rear; but their flesh is certainly very delicate, and perhaps more so than that of other kinds—a circumstance, however, that may partly result from their far greater delicacy of constitution, and the consequent extra trouble devoted to their management.

The brown and ashy-grey are not particularly remarkable; but the black are decidedly superior in every respect, not only as regards greater hardiness, and a consequent greater facility of rearing, but as acquiring flesh more readily, and that being of the very best and primest quality. Those of this color appear to be less far removed than the others from the original wild stock. Fortunately, too, the black seems to be the favorite color of nature, and black turkeys are produced far more abundantly than those of any other hue. M. Parmentier was informed by a French lady, who had devoted much of her attention to rural affairs, that she had in her yard ten black turkey hens and a white cock, and yet, that not one of the chicks was white, or even light-colored. Turkeys will sometimes change their hue. Mowbray states that "A turkey cock, which was black in the year 1821, became afterwards perfectly white, this extraordinary change taking place so gradually, that in the middle of the moulting the bird was beautifully mottled, the feathers being black and white alternately."

With respect to the best mode of keeping turkeys, I have merely to repeat what I have already remarked relative to a due attention to the habits of the original wild breed in its native state. Let them have a large, roomy, open shed, sufficiently protected, of course, from the weather, and, above all, from moisture. Let the perches be high—and here, again, you will do well not to omit the use of the hen ladder; for, although these birds can usually fly well, still, when fat, they become too heavy for their wings, and are apt to injure themselves in their descent from a lofty perch, especially when in confinement: when at full liberty they can take better care of themselves. During warm weather they may be permitted to select their own roosting-places on the trees about a farm; but should be well watched, lest they stray away; and this indulgence should on no account be granted them if frost be anticipated, as their toes are tender and apt to become frost-bitten. Indeed summer is the only time of the year when this out-roosting may, with safety, be permitted.

The turkey is a profitable bird, for it can almost wholly provide for itself about the roads and hedge-rows: snails, slugs, and worms are among the number of its dainties, and the nearest stream serves to slake its thirst. To the farmer, however, it is often a perfect nuisance, from its love of grain; and should, therefore, be kept in the yard until all grain is too strong in the root to present any temptations.

Notwithstanding the separation which, with the exception of certain seasons, subsists, in a wild state, between the cock and hen turkey, they have been brought to feed and live amicably together in a state of domesticity. The former, however, retains sufficient of his hereditary propensities to give an occasional sly blow to a chick, or forward poult, but that very seldom of a seriously malicious character.

Mascal', in describing a turkey cock (such as the breeder should select) says, that he should be a "a bird large, stout, proud, and majestical; for when he walketh dejected, he is never good."

M. Parmentier says that both cock and hen should have short legs, full shapes, and general vivacity and energy in all their movements; likewise, that they should be both well shaped and in healthy condition.

Mascall says, that the cock should not be "passing a yere or two yeres old: three yeres is the most, and too much."

For my own part, I hold a turkey cock, at the age of three years, to be only in his prime, and to continue, in every respect, suitable for your purpose until five. The hen is at her prime younger, and, probably, at the second year is as good as ever she will be afterwards.

It has been stated by some, and yet as positively denied by others, that one fecundation will render all the eggs of that laying fertile; still, however, were it my own case, I should prefer making "assurance doubly sure," by allowing one cock to every dozen or fourteen hens.

The approach of the laying season is easily known by the increased liveliness and proud strut of the hen; and she likewise further expresses her feelings by a peculiar self-satisfied cry, that soon becomes familiar to the observer. This usually takes place in the month of March (nearly a month earlier than with the wild bird). When the breeder perceives these symptoms, he should provide a nest, and put an egg, or a bit of chalk formed like one, into it, to induce the hen to commence laying there. Partaking of the retiring propensities of the wild hen (although she has not equal reason to dread the destructive passions of the cock), the turkey is a secret layer, and does her best to elude the vigilance of her keeper and steal away to some secluded spot. The peculiar note of which I have spoken, betrays, however, the fact; and whoever has the care of the fowl, should trace her to her retirement, and bring her back to the nest prepared for her.

The time when the hen turkey lays is usually morning. Some lay daily; others only every second day. The number of eggs laid is commonly from fifteen to twenty; but this varies with the age of the bird, a hen of mature age laying more and larger eggs than one of a year old. When the turkeys' are to be let out in the morning, you may examine the hens, and keep in such as are about to lay. This precaution will, of course, prevent the loss of a single egg. When the hen is laying, the cock should be kept from her, as he would ill-treat her and break the eggs. The eggs should be taken away as soon as laid, lest they might be broken through the awkwardness of the hen, or sucked by vermin. They will keep till the hens are done laying, if put in a basket and hung up in a dry place. It is unnecessary to keep the eggs belonging to each in a separate place. The hen turkey is not troubled with any very exclusive feelings, or, rather, her disposition overflows with an excess of maternal love: for she will rear a brood belonging to another quite as carefully as if they were her own. In the second laying, the eggs are fewer in number, seldom exceeding from ten to thirteen; and on this occasion extra care is requisite.

The sooner that one hen is turned away from her brood, and the brood mixed up with that of another, hatched about the same time, the better chance there is of rearing it, as the hen which is so turned away, will lay again in a fortnight or three weeks, and thus hatch a second time before the month of July is out. Even under these circumstances, the chance of rearing the young ones is very uncertain, as they are hardly strong enough to meet the cold nights in autumn, when they often become what is called *clubfooted*, and die. I rather recommend letting the hen lay as many eggs as she will, and turning her off when she becomes broody. Hens thus treated will lay again in the month of August; so that, under all circumstances, they may be called profitable birds.

The turkey hen is a most persevering sitter; and when her eggs are taken away, she would sit upon stones, if she could not procure the eggs of another bird, and would perish before quitting the nest. Eggs should therefore, be left with her, not only to tranquillize her, but because sitting upon eggs fatigues her less than sitting upon an empty nest; but these eggs must be marked in order to distinguish them from those the poor bird continues to lay; for any eggs that seem to her to be slow of hatching, will be abandoned, as she will quit the nest as soon as she perceives the chick; consequently, as soon as the eggs you have placed under her are hatched, she will leave the nest, and the eggs of her own laying will be sacrificed. Remove, therefore, the former; and it is for this reason that I recommend them to be marked. Keep the nest clean while the turkey hen is sitting, as dirt will injure the eggs. No one should go near a hen when sitting, except her keeper; and no one should turn the eggs, or meddle with them further than I have already indicated. The bird will turn her eggs with more judgment than you can do.

On the thirty-first day of sitting, the chicks leave the eggs; but as some quit their prison before others, they must be placed in a basket filled with feathers, and if the weather be cold, placed in some warm spot. When all are out, they may be given to the hen, for six or eight hours before feeding. Sometimes the chick will require assistance in leaving the egg; and, if so, the same caution must be observed that I have insisted upon in the case of the common fowl. Be very sparing of your aid, or you may do far more harm than good.

Many writers recommend a vast deal of quackery in the treatment of the young chicks. Some go the length of ordering them wine, pepper, bathing in cold water! &c. It is far better to let them alone. For a few hours after hatching, the chicks require no food at all; and then, instead of cramming them—a process in which you are likely to break the tender beak of the little chick—chop up a few hard eggs with boiled nettles, parsley, and a little bread or curd; make this into a paste, and present it to the birds in the palm of your hand, or place it before them on a stone, taking care that the hen does not rob them. In supplying them with water, be careful to put it in such very shallow vessels that they cannot wet themselves; for the least moisture appears fatal to them. As the turkey chick does not seek its food immediately on leaving the egg, as the hen seems incapable of instructing her little offspring how to do so, it is a practice with some to put a few common hen's eggs among the turkey's, (which must be done about nine or ten days after sitting), that these, coming out with the little turkeys, may, by force of example, teach them to provide for themselves.

Unless in very warm weather the hen and chicks should be housed for a month. If they appear drooping, put powdered caraway seed, and a little Cayenne pepper into the food. If you mix the food with milk, let it be previously boiled. Unboiled milk will purge the chicks; but, for my own part, I prefer pure water.

At the age of about two months occurs the most critical period in the life of a turkey, called "shooting the red;" or the time when the head and neck acquire the reddish color of the adult. This crisis once past, the birds may be regarded as past danger, and exchange the name of chicks for that of turkey poults. The only treatment necessary when the bird is shooting the red is to furnish nutritive food, with the addition of a small pinch of Cayenne pepper. Bruised hempseed is also found serviceable.

I know no birds better calculated to be profitable to the breeder, than turkeys. They will almost wholly provide themselves with food; and it is only the young chicks that require nourishment at our hands; and how inexpensive, also, is this nourishment! With care you may rear two broods in a year, and have from eight to fifteen survivors in each. Take the average at ten, and, suppose you have three hens, you may bring up thirty chicks. Your hens will cost you nothing for keep; and you must admit that your profit is handsome. This is, however, far below the mark. There is nothing to prevent an individual having more hens, rearing larger broods.

Cobbett, a shrewd and accurate observer, thus writes—" As to the feeding of them when young, many nice things are recommended—hard eggs, chopped fine, with crumbs of bread, and a great many other things; but that which I have seen used, and always with success, and for all sorts of young poultry, is milk turned into curds. This is the food for young poultry of all sorts. Some should be made fresh every day; and if this be done, and the turkeys be kept warm, not one out of a score will die. When they get to be strong they may have meal and grain; but still, they always love the curds. When they get their head feathers, they are hardy enough; and what they then want is room to prowl about. It is best to breed them under a common hen, because she does not ramble like a hen turkey; and it is a very curious thing that the turkeys bred up by a hen of the common fowl do not themselves ramble much when they get old.

"The hen should be fed exceedingly well, too, while she is sitting, and after she has hatched; for no man ever saw healthy chickens with a poor hen. This is a matter much too little thought of in the rearing of poultry; but it is a matter of the greatest consequence. Never let a poor hen sit; feed the hen while she is sitting, and feed her most abundantly while she has young ones, for then her labor is very great. She is making exertions of some sort or other during the whole twenty-four hours; she has no rest; is constantly doing something, in order to provide food or safety for her young ones. As to fattening turkeys, the best way is never to let them be poor. Cramming is a nasty thing, and quite unnecessary. Meal mixed with skim-milk, given to them fresh, will make them fat in a short time, either in a coop, in a house, or running about. Boiled carrots and Swedish turnips will help, and it is a change of sweet food."

As observed already, once the turkey chicks *shoot the red* (which takes place at or about eight weeks old), they may be considered out of danger; hence, many persons consider it more profitable to buy lean young poults, after they have got the red, and then fatten them for market, to breeding them. If the mortality among the chicks were greater, and were not so easily to be avoided by a very little care, this might be the preferable mode of going about the matter; but as it is, there can be no doubt of the greater advantage to be derived from rearing your own chicks.

In feeding the poults, after the second month, it will suffice to give them such boiled common plants and herbs as are of a nutritive character—nettles, wild succory, milfoil, turnip tops, cabbage sprouts, or the outside leaves of greens well boiled down—with these, potato skins and an odd potato or two itself may be given, and acorns, if they can be had without expense. The meal of buckwheat, barley, beans, oats, according to whichever is most plenty with you, will, when incorporated as I have described with potatoes, fatten the poults with great rapidity. But, you may also use the meal of Indian corn with the greatest advantage, though it requires treble the boiling of oatmeal. If you desire to meet the market hastily, and with profit, you will be compelled to resort to more expensive feeding than otherwise, but you will be repaid by the result. When the poults are about five months old, or earlier, if it be late in the season and cold weather seems at hand, give them boiled potatoes mashed with meal, and then chopped small, as I have described. Let this be given fresh, and the vessel in which they are fed well washed daily, as otherwise it will speedily contract a sour smell and become repulsive to the birds, for turkeys are both cleanly and nice in their appetite. After having persevered in this feeding, morning and evening, for about a month, during which time the exercise of the poults should be greatly curtailed, and they should likewise be kept much of their time (especially after meals) in the dark, they will be found fit for use, and, if of a good kind, at least eighteen pounds weight.

As damp or cold is fatal to turkey poults, so is intense sunshine. Poults should not be suffered to stray too far; for, independently of the risk they incur, in case of sudden shower, it must be remembered that they are as yet incapable of encountering any great fatigue, and that their condition will be anything but benefited thereby. Mr. Dixon recommends a mode of diet that I have never seen tried:—" No food makes their flesh whiter and more delicate than kitchen stuff, or the dregs of melted tallow, more or less of which must be boiled according to the number that is to be fed; and being diluted in a boiling kettle, plants (and especially nettles chopped up) and pot-herbs are mixed with it. The whole being well boiled, barley-meal or corn is added to form a kind of paste, which may be given twice a day at least— in the morning and at one o'clock—when it is wished to render them fat. But as the dregs of melted tallow are not everywhere to be procured, the dregs or refuse of the oil of nuts, linseed, or sweet almonds, may be substituted, the greatest care being taken not to fatten them wholly with such oily substances, for their flesh would partake of the flavor and be injured."

From what we know of the value of oil-cake in the fattening of our cattle, I have no doubt of its efficacy in fattening turkeys, but it certainly renders the flesh rank and oily. In reckoning the advantages with the expense attendant on the rearing of these birds, until you want to fatten them for sale or your own consumption, you need be at no pains relative to their food, as they are quite able to provide for themselves, being in this respect superior to any other of our domestic fowl. In thus readily providing for themselves, they are also greatly assisted by the easy character of their appetite—grass, herbs, corn, berries, fruit, insects, and reptiles; in short, hardly anything coming amiss to them.

Audubon says, that in their native forests "they cannot be said to confine themselves to any particular kind of food, although they seem to prefer the peccan nut and winter grape to any other; and where these foods abound, are found in the greatest numbers. They eat grass and herbs of various kinds—corn, berries, and fruits of all descriptions. I have even found beetles, tadpoles, and small lizards in their crops."—*Ornith. Biog.* l. ii. A favorite repast of this bird in its native forests is said also to be the seed of a kind of nettle, and at another season a small red acorn, on which latter food they soon become so fat that they cannot fly, and are easily run down by dogs.

They are dull and stupid at getting the corn out of the ear, unless very ripe, and will walk through a field of peas or beans without opening a single shell, even if they are ripe.

There are many sorts of food which, though nutritious and highly salutary as concerns other fowl, are little short of downright poison to turkeys. Amongst others, I may enumerate vetches or tares, marrowfat peas, and most sorts of pulse.

The weight of turkeys has been much exaggerated by careless, ignorant, or, perhaps, credulous writers; and 60 lbs. is, by some, mentioned as a common weight. On the contrary, 20 lbs. is a fair weight for any fat yearling bird (and a very great weight for a bird of six months old); 30 lbs. is a fine turkey of any age; and few ever exceed 40 lbs. The greatest weight that these have been known to attain, recorded by such authority as we can rely upon, is 56 lbs. I have never seen a turkey of 60 lbs. weight: nor do I know any one that has. The hen takes fat more readily than the cock, and is, in proportion to her size, a tenderer and better dish.

THE GUINEA HEN, OR PINTADO.

The original country of the Guinea fowl is, as its name implies, Africa; but it is likewise common in America, where it is supposed to be indigenous, as well as the turkey.

THE GUINEA FOWL.

The Guinea fowl is slightly larger than the ordinary barn-door fowl, but is inferior in size to the larger foreign breeds, as the Malay and Spanish; in both aspect and character it appears to occupy a position between the pheasant and the turkey. Although long familiarized, the Guinea fowl has never been fully domesticated, still retaining much of the restlessness and shyness of its primitive feral habits. It is very courageous, and will not only frequently attack the turkey, but even prove victorious in the encounter.

The cock and hen are so nearly alike, that it is not easy to distinguish them; there is sometimes a difference of hue in certain parts; but this difference only occurs occasionally, and indeed, it is on gait, voice, and demeanor, that we must chiefly depend. It must be remarked that they pair; therefore a second hen will be neglected and useless except for eggs.

As a source of profit I cannot recommend these fowl: the eggs are very small, and the flesh not being likely to please every palate. Still, however, as the Guinea fowl require but little trouble or attention, and their eggs, though of small size, are well flavored and numerous, they are generally kept wherever there is accommodation for them. The chief objection to them is their cry, or scream; and even this, again, has its advantages, invariably predicting a change of weather: they can hardly, however, be kept with other poultry, on account of their pugnacity.

The Guinea fowl dislikes confinement, and will not thrive unless it has free liberty; where such, therefore, cannot be afforded, it is useless to attempt keeping it.

These fowl are prolific; the hen commences to lay in May, and lays throughout the entire summer. The period of incubation is twenty-eight days; but it is more advisable to keep the Guinea hen entirely for laying, and if you desire to hatch any of the eggs, to do so under the hen of the common fowl. You must keep the male bird away, or he will destroy the eggs.

The chicks, while young, require careful management, and must be constantly fed; in a short time they become perfectly hardy. At nine months they are fit for the table.

PEA-FOWL.

A Peacock in full feather, parading on a green lawn, or from the extremity of a terrace-wall, displaying the full length of his gorgeous tail, is one of the most beautiful of living additions to garden landscape. But of fruit he will prove a devourer, not to be guarded against, and both he and his mate are not unfrequently murderous assassins of the young of other fowl. The cock does not attain the full splendor of his plumage until he is three years old, and the hen does not lay until the same age. She lays from five to seven eggs, and sits twenty-nine days. If the first batch of eggs be taken away, she will lay a second, so that by having a hen turkey foster nurse you may manage to have two broods in one summer. The peahen generally chooses a very retired spot, quite out of the way of the peacock, who is often a cruel unnatural father. The young must be hatched like Guinea fowl and young turkeys: unless amply and regularly fed they are apt to wander. When fat and hung long enough, they make a delicious and splendid roast. They should be larded with slices of fat bacon, the head and neck with the feathers on, carefully wrapped in paper, and tucked under the wing away from the fire, and when ready set up in purple glory, to match the tail adorned with feathers, neatly stuck in at the last moment. If you wish pea-fowl to agree with other poultry, they must be reared with them.

CHAPTER IX.

WEB-FOOTED BIRDS.

(ANATIDÆ).

A FAMILY of web-footed birds whose habits are, generally speaking, aquatic, though some of them are more so than others. This family of the Palmipedes of Cuvier have a large and broad bill, the edges of which are beset with *laminæ* placed transversely. They are divided into Swans, Geese, and Ducks. The limits of each, however, are not very well defined.

THE SWAN.

Swans (Cygnus) are found on the rivers and small pools of fresh water, rather than on the sea or the larger lakes, and, when they do appear on these, they are always found near the shores, and never on the expanse of the broad waters. The chief reason of this is, that they are vegetable feeders, and although their long necks enable them to reach the bottom at considerable depths, they never dive, and they rarely feed upon the land, or in any other mode than by floating on the surface of the water. They are among the most ornamental of all the water birds, on account of their great size, the gracefulness of their forms and motions, and the snowy whiteness of the plumage of those species with which we are most familiar. Swans have, from the remotest antiquity, attracted the attention of poets and other describers, and the ancient fable of their acquiring a musical song when they are dying, instead of the husky voice which they have when alive, is still repeated, though wholly destitute of foundation.

In some of the species, the swans approach the geese in many of their characters, while the typical ones differ considerably.

The Mute or Tame Swan (Cygnus olor) is "the Swan," by way of eminence. The bill of the mute swan is of a red or salmon color, with the margins and the basal cere, which swells into a tubercle of considerable size, black; the whole plumage of the mature bird, when on the water in a pure atmosphere, is beautifully white; and few of the living productions of nature are more beautiful than swans, especially when they are upon small expanses of clear water. Though a majestic creature in its motion upon the water, the appearance of the swan harmonizes best with that which is clear and tranquil, and grasses and green meadows add greatly to the effect.

In a state of nature this species is migratory, where the seasons run into extremes; when wild they are compelled to move southward when the weather is severe; but where the waters are open they continue on the same grounds for the whole year round, and where they are placed upon ornamental waters in pleasure-grounds, or even in the close vicinity of cities, they show no very strong disposition to shift to more sequestered haunts, at any season of the year. In places that are much frequented they soon become very familiar; indeed they are far from being timid birds under any circumstances. They appear to be quite confident in the power which nature has given them; and, as they have little to fear from enemies, they are not much given to be pugnacious, at least in ordinary times of the year. When, however, they have nests, they not only defend them with great bravery, but attack in the most resolute manner, any animal that approaches, not excepting man himself. The female is a close sitter during her incubation, which is forty days; and while the female sits, the male is very assiduous in watching for the safety of the family. He is ready to resist, and by the most vigorous means to repel, every intruder, not excepting his own species, who cannot come within a short distance of the nest without being attacked. Severe contests often take place between the males upon these occasions, more especially if, as is sometimes the case, there is an odd or unpaired male upon the same water. This odd one is not the assailant; for, as he is not in the guardianship of a female and nest, he does not appear to have the same excitement as those which have this trust committed to them; but if he is attacked, he is bold enough in self-defense; and it has been stated that if he should succeed in killing or beating off the legitimate possessor of the ground, even after the incubation is considerably advanced, he takes the place and discharges the duties of watchman and protector, with the same vigilant assiduity as the one which he has vanquished.

The Swan forms one of the finest ornaments of a sufficiently extensive sheet of water, and a pair will keep down weeds much more cheaply and effectually than any mechanical appliance. An island will be found the best

DOMESTIC SWAN.

oreeding place. They require feeding during winter, at least; but, it is better to feed them constantly. A fat young cygnet affords a delicious dish. Swans, as well as all kinds of wild or semi-wild water fowl, must be pinioned, or they will be apt to depart without leave at the improving period of the year. To effect this operation, find the joint of the bastard wing, which will include about five flight feathers, introduce a sharp knife between the joints, cut steadily and boldly: no injury will ensue. The Swan begins to lay at three years old.

The nest of the SWAN IN A DOMESTIC STATE is large and wide; the eggs are of a white color, and vary from six to eight in number. The cygnets are grey, and do not acquire their full plumage till the second year, and till then they usually keep in company with each other, which they also do with the old birds, until the time of pairing again comes on. The cygnets while they are in their grey plumage, have very little of the majestic appearance of the adult swans. As articles of food, they are, however, the only ones that are held in much estimation, and there is probably more of the want of rarity than that of nature in them. Taken from the water in their natural condition, they are comparatively of little value; but, when they are artificially fattened, they fetch a high price in the market. When tame, swans are kept with a view to profit as well as ornament; their down and the quills of their wings are pulled twice in the year. This is a very cruel operation; but then, the feathers pulled from the live bird are better than if they were taken from it when dead; and, if the operation is performed near the time of the moult, and the birds are well fed, it is not so hurtful to them as might at first be supposed.

The Black Swan (C. niger,) which is a native of Australia, but has been domesticated. It is much more of a tyrant on the waters than the White Swan, and will allow no other swimming bird to live in its vicinity. The whole plumage is black, with the exception of the first six quills, which are white; the bill, and a naked space round the eye, are red; the length is about four feet and a half, and the wings rather shorter in proportion than the White Swan, but they are broad and strong. The plan and structure of the nest are about the same as those of the White Swan, and there does not appear to be much difference either in the food or the general habits. The male is particularly watchful of the female when sitting, and of both female and brood when they are on the water; he not only drives off all other birds, but if any animal, or even a human being approaches, he lands and marches forth to give him battle at a distance from the family; his wings are raised ready for the stroke, his feathers ruffled, and he puts on altogether rather a formidable appearance. The young are produced about the same season as those of the White Swan, and the number in a brood appear also to be the same. They are of a blackish ashen grey, which continues the whole of the first year. As a curiosity the Black Swan is very well, the more especially that it was for such a length of time implicitly looked upon as the impossible bird that was nowhere to be met with; but it has none of the beauty and grace of the White Swan, which must continue to be the favorite as an ornamental bird.

THE GOOSE (ANSER).

Geese are very numerous, as well in species as in varieties. They are more abundant in the polar countries than in the southern regions; and, with few exceptions, are completely web-footed, and can swim. Swimming is not, however, their proper and peculiar, or, in general, even their chief motion. If the structure of a goose, and the way in which the legs support the body, are compared with those of a duck, we shall perceive a very remarkable difference in the purposes for which they are best adapted. The bodies of ducks are "boat-built," and evidently formed for getting through the water rapidly at a small expense of effort; their legs are placed far backward, so as to strike against the water which follows in their wake; while the Goose is properly a walker, although the power of swimming is added, and in some of the species the two powers are nearly equal, while there may be some in which the swimming predominates.

Geese are also much more exclusively vegetable feeders than the rest of the *Anatidæ*; at least, with the exception of the swans, which are also much more aquatic in their feeding than the geese, for which habit they are well adapted by the greater length of their necks. Geese never dive, nor do they, in many instances, feed below the surface of the water, though they often feed, while swimming, on the seeds and succulent leaves of floating aquatic plants.

The generic characters are: the bill shorter than the head, higher than wide at the base, diminishing towards the tip, and thus having a slightly conical form. The teeth, in the margins and toward the tip of the bill, are conical, and the point of the upper mandible is generally furnished with a nail of harder consistence than the rest, and sometimes differently colored. They are, generally speaking, polygamous; but there is no great external difference between the sexes. The old males are, indeed, rather larger than the females; but, before they reach maturity the two sexes are very much alike both in size and color.

The natural habitats of the geese are damp meadows, and tufted marshes which abound with plants, a species of pasture which naturally points out why geese in a state of nature should be very migratory birds.

The goose is a bird of no mean consequence in history. The Roman geese gave warning of the approach of the foe, and saved the Capitol; and it is from this circumstance, according to some, that this bird has since been a favorite Christmas dish. On account of this valuable service rendered by the goose to the Roman state, it had the honor of being eaten with great pomp at important public festivals; amongst which were the *Julian* games. The Romans introduced the goose into Britain; *Yule*, the Scotch term for Christmas, is derived from *Julius*, and hence the goose is a Christmas dish. I confess I think this a little far-fetched, and suspect the true reason to be, that at that period of the year the goose is in the best condition, and fittest for the table.

THE COMMON WILD, OR GREY-LAG GOOSE (*A. palustris*).—This is generally understood to be the parent stock of all the domestic species of Europe.

The Gray-lag is about thirty-five inches in length, the female being somewhat smaller. Its beak is of a pale flesh color, with the nail, or horny tip, white; the iris is brown; the head and neck of an ashy gray; the inner part of the wings pale leaden gray; belly and under surface of neck white; legs of a very pale flesh color. The chief characteristics of the Grey-lag are the light ashy-blue color of the outer portion of the wing, and the conspicuous white extremity of the beak. The Gray-lag wild goose is now rarely to be met with.

THE WHITE-FRONTED GOOSE.—The length is about two feet four inches, the extent of the wings about four feet and a half, and the weight about five pounds. The bill is thick at the base, of a yellowish red color, with the nail white. A white patch is extended over the forehead from the base of the bill and corners of the mouth. The rest of the head, neck, and the upper parts of the plumage in some specimens are dark brown, and each feather is margined more or less with that color; the primary and secondary quill feathers are of the same, but much darker, and the wing-coverts are tinged with ash. The breast and belly are dirty white, barred with irregular patches of very dark brown, and tipped with lighter shades of the same color. The tail is horny ash-colored brown, and surrounded with white at the base; the legs yellow.

Of these four varieties, the Gray-lag and the White-fronted are obviously the originals of our domestic geese. The legs of many of our domestic geese are orange-colored, like those of the White-fronted. The legs of the wild Gray-lag goose are of a pale flesh color.

THE CANADA OR AMERICAN WILD GOOSE.—In the slenderness of its make, and the form of its neck, this bird somewhat approaches the swan. The black and wing coverts are dull brown, each feather having a whitish tip; side pale ashy brown; upper part of head and neck black, with a broad patch of white spreading from the throat over the lower part of cheeks on each side; the bill is black; legs and feet grayish black. This bird is easily naturalized amongst us, and affords good flesh for the table; in captivity it readily pairs with the common gray goose, and the young are superior to either parent in point of size. The principal objection to the breeding of the Canada goose as a member of our poultry establishment, is its not being prolific, and hence not affording promise of being profitable.

CANADA OR AMERICAN WILD GOOSE.

DOMESTIC GEESE, AND THEIR MANAGEMENT.

Amongst the varieties of our common domestic goose we must first describe one which, though of comparatively recent introduction, and as yet not generally to be met with as an ordinary inhabitant of farm-yards, bids fair, from its unusual size, and capacity of carrying flesh, shortly to supersede every other in the estimation of the fancier or breeder. This is

THE TOULOUSE GOOSE.

This bird was originally imported from the Mediterranean, and is known indiscriminately by the names of Mediterranean, Pyrenean, or that of Toulouse. This bird is chiefly remarkable for its vast size—a property in which it casts every other known breed far into the shade; it is indeed, the MAMMOTH of geese, and it is to be regarded as a most valuable addition to our stock. The prevailing color of the Toulouse goose is a slaty blue, marked with brown bars, and occasionally relieved with black; the head, neck, as far as the beginning of the breast, and the back of the neck, as far as the shoulders, of a dark brown; the breast is slaty blue; the belly is gray, as also the under surface of the tail; the bill is orange red, and the feet are flesh color. There can be little doubt of this valuable bird being the unmixed and immediate descendant of the Gray-lag, and it was, indeed, at once pronounced to be such by the Royal Zoological Society of London, at their poultry exhibition of June, 1845.

In habit the Toulouse goose resembles his congeners, but appears to possess a milder and more easy disposition, which, greatly conduces to the chance of his early fattening, and that, also, at little cost. Of his other peculiarities, the curl of plumage on the neck comes closer to the head than in the common geese, and the abdominal *pouch*, which, in other varieties is attendant only upon age, exists in these birds from the shell; the flesh of the Toulouse goose is tender and well flavored. As a cross with our common domestic goose, I am certain they will be found most valuable, and we may thus expect eventually to arrive at a degree of perfection not hitherto anticipated. Mr. Dixon considers this to be no species or variety, but merely a well grown specimen of the common goose, raised in warm weather, and amply fed, and he is probably correct; as I have reason to believe that we diminish the size of our geese, and other poultry, by killing them off before their maturity.

Several other species of southern geese are mentioned as being found on the Falkland Islands, on Terra del Fuego, and some other places of the southern lands; there have also been others brought from South America; but all these are too little known, we are too little acquainted with the migration of birds in the southern hemisphere, and those migrations are in themselves on so small a scale compared with the migrations in the north, that all that could be said about those birds would be little else than a description of colors. There are, however, some other species which require a brief notice, because they deviate in some respects from the typical characters of the genus.

These species which deviate from the proper character of the geese in many points, but which still essentially retain that character in others, may be divided into two sections: First, those which form a sort of intermediate link between the geese and the swans; and secondly, those which form a similar link between the geese and the wading birds, more especially the Crane family, or perhaps the Herons. We shall take them in the order now stated, without being very particular as to the correctness of the names, because, though we are not quite satisfied with the existing ones, we do not feel ourselves called upon to contrive new ones, as our object is not to make systems but to give useful information.

THE CHINESE GOOSE (*A. Cygnoides*). This species is not called cygnoides, or swan-like, from any actual resemblance that it has to a swan in any other respect than in color; and that is not constant, for though it is sometimes entirely white, it is subject to great variety of shade. Though specimens have been brought from China, it is perhaps not very correctly styled the Chinese goose, inasmuch as it is found in many other parts of the south-eastern world, from China to the Cape of Good Hope, and it is said, from New Zealand, though it does not appear to be met with in New Holland.

THE CHINESE GOOSE.

In addition to the Chinese Goose, already described, there are three sub-varieties, each presenting striking points of difference, and yet being sufficiently alike to justify me in classing them together. These are—

1. THE HONG KONG.—This bird has a large horny knob on the bill and forehead; its prevailing color is gray, with a longitudinal stripe of a deep brown running above the back of the neck. The legs are of a red color, whence it is sometimes distinguished as the "Red-legged China goose." This is the same long known amongst us under the erroneous name of the "Poland goose."

2. THE BLACK-LEGGED CHINESE GOOSE. Also knobbed, and usually with a white edging round the knob, somewhat similar to that of the wild breed called the "White-fronted goose."

3. THE WHITE CHINESE GOOSE. A very handsome bird, knobbed as the rest, of a snow-white color, and with legs of a bright orange red.

These geese are inferior in size to the Toulouse, but nevertheless very fine birds, and worthy the attention of the breeder. The white variety, especially, with red legs, is very beautiful, and would form an appropriate ornament on a piece of water. The flesh of the Chinese goose is also good; they feed well, fatten easily, and are very prolific.

THE COMMON DOMESTIC GOOSE.

Of our ordinary and well known domestic geese there exist but two sorts, whose only distinction seems to rest in their relative size, they being divided into the large and small; and by some, according to their color, into the white and the gray. These divisions are, to a certain extent, arbitrary; as out of one brood you will generally find the several varieties, both as to size and color, that you seek. I may, however, inform the reader that the best sorts of geese are those which vary least in color. Those approaching most nearly to the primitive stock, are the birds which every best judge will prefer breeding from. *Gray* is the best color as coming nearest to the original Gray-lag; white is not quite so good; but avoid mixed colors; they will not prove so prolific, and the young will be more difficult to feed up to the required standard.

I have heard it recommended to try the experiment of crossing with the original wild stock. This would, no doubt, be a most excellent plan. The Gray-lag would be the mark; but it is now scarce. This was doubtless the plan resorted to by the Spaniards, whom we have to thank for our recent invaluable acquisition of the Toulouse variety. All we have to do now is to avail ourselves, as far as possible, of the superb cross thus brought within our reach; and we may, ere long, bring up our common breed of gray geese to equal that of the continent.

As to breeding geese. These birds, as has been ascertained by M. St. Genis, will pair like pigeons; and even if the number of ganders exceed that of the geese, no noise or riot takes place, mutual choice being evidently the ruling principle. Amongst other experiments tried by M. St. Genis, he left, besides the patriarch of the flock, two of the young ganders, unprovided with mates, but still those couples that had paired kept constantly together, and the three single ganders never attempted to approach any of the females during the temporary absence of their lords. M. St. Genis also remarked, in the course of his observations, that the gander is more frequently white than the goose.

Some writers recommend a gander to be mated with from four to six geese. As I have already remarked, when treating of poultry, this must entirely depend on the object the breeder may have in view. If he desire eggs, and eggs alone, one gander is plenty for six or even eight geese. He may, indeed, abandon the unnecessary trouble of keeping a gander at all. It, however, so happens, that keeping geese for the produce of their eggs alone, is anything but profitable; and hence these must be rendered duly fertile; and, to effect this, one gander to an almost indefinite number of geese will not answer. For the purpose of hatching, a gander should be mated with, at most, four geese. Let him be, if of the ordinary kinds, amongst which color varies, of a pure white or ash-gray color; but not at all of two colors. Let his size be large, his gait active, his eye lively and clear, his voice ever ready and hoarse, and his demeanor full of boldness and impudence. Select the goose for her weight of body, steadiness of deportment, and breadth of foot—a quality that, however unfeminine it appear, happens, in the instance of geese, to indicate the presence of such other excellence as we require.

The goose deposits from ten to twenty eggs at one laying; but, if you do not desire her to sit, you may, by removing the eggs as fast as they are laid, and, at the same time, feeding her highly, induce her to lay on from forty-five to fifty. This is, however, unusual, and it is unprofitable. When tolerably well cared for, geese may be made to lay, and even hatch, three times in the year. This care consists merely in high feeding and good housing early in the spring, so as to have the first brood early in March; but I would rather have two good broods reared than three bad ones, and I am, therefore, more disposed to recommend patience and moderation.

The goose will, when left to the unassisted promptings of nature, begin to lay about the latter end of February, or the beginning of March. The commencement of the laying may be readily foreseen by marking such geese as run about carrying straws in their mouth. This is for the purpose of forming their nest, and these individuals are about to lay. They should, then, of course, be watched, lest they drop their eggs abroad. Once a goose is shut up, and compelled to lay her first egg of that laying in any particular nest, you need be at no further trouble about her; for she will continue to lay in that spot, and will not stray on any account elsewhere.

We can always detect the inclination of the goose to set or hatch. This is known by the bird keeping in the nest after the laying of each egg longer than usual. The hatching nest should be formed of straw, with a little hay as a lining; and so formed that the goose will not fling the eggs over the side when in the act of turning them. You need not banish the gander; on the contrary, let him remain as near the nest as he chooses: he will do no mischief, but will act the part of a most vigilant guardian. About fifteen eggs will be found as many as a good-sized goose can properly cover. Do not meddle with the eggs during the incubation, and do not meddle with the goose; but, as she is somewhat heavier than the hen, you may leave her food and drink rather nearer to her than is necessary with common poultry, as, if she chanced to absent herself from the eggs sufficiently long to permit them to cool, she might become disheartened, and desert her task altogether. It is, however, unnecessary to put either vinegar or pepper in her food or water, as recommended by some, or, in short, to meddle with her at all.

The goose will sit on her eggs for nearly two months; but the necessary period of incubation being but one, the early hatched goslings must be removed lest the more tardy might be deserted. About the twenty-ninth day the goslings begin to chip the shell; and if their own powers prove inadequate to their liberation, aid may be rendered them, and that, also, with much less risk than in the case of other young birds, the shell and its membranes being very hard and strong, and the young themselves also hardy, and capable early of enduring hardship. The best plan is to have the eggs set, of as nearly as possible equal freshness, that they may be hatched at the one time.

On first being hatched, turn the goslings out into a sunny walk, if the weather will permit of such procedure; but do not try to make them feed for, at least, twelve hours after leaving the shell. Their food may then be bread soaked in milk, porridge, curds, boiled greens, or even bran, mixed with boiled potatoes, taking care not to give the food in too hot a state, while you equally avoid giving it cold. Avoid rain or cold breezes; and see, therefore, that the walk into which you turn the young goslings be sheltered from both wind and weather. The goslings should also be kept from water for at least a couple of days after hatching. If suffered too early to have free access to water, they are very liable to take cramp—a disease which generally produces permanent lameness and deformity, and but too frequently proves fatal.

Geese should have an enclosed court or yard, with houses in which they may be shut when occasion requires. It is better, however, to confine them as little as possible; and, by suffering them to stroll about, and forage for themselves, the expense of rearing them will fall comparatively lightly on you, so that you will not be conscious of any outlay. Geese require water, and cannot be advantageously kept when they cannot have access to it; still, however, I have known them thrive where they had no access to any pond or river, but had only a small artificial pool, constructed by their owners, in which to bathe themselves. When geese are at all within reach of water, they will, when suffered to roam at liberty, usually go in search of, and discover it, and will, afterwards, daily resort thither. Though the birds are thus fond of water, all damp about their sleeping places must be scrupulously guarded against. Grass is as necessary to the well-being of geese as water; and the rankest, coarsest grasses, such as are rejected by cattle, constitute the goose's delicacy. Such grasses as they prefer, will be found on damp, swampy lands, of which, perhaps, no more profitable use could be made. The stubble-field is, in its season, an excellent walk for geese; for they there not only find the young grass and other herbage springing up amongst the stubbles, but likewise pick up much corn that would otherwise be lost. When the stubble-field is not to be had, there is usually something in the kitchen-garden that would be wasted if the geese were not turned in; and, observe, that this is the only season when these birds can be suffered to enter a garden; for they are very destructive both to farm and garden crops, and even to young trees. Geese do not answer to be wholly fed on such green food as they can provide for themselves; but if they get a few boiled potatoes occasionally, bruised up with a little bran, and not given too warm, they will be raised for the market at scarcely any cost, and will, consequently, be found very profitable to the farmer. Market gardeners should never be without geese, which would consume all their refuse, and bring money into their pockets, in return for their consumption of what would otherwise be wasted.

Various measures have been adopted for fattening geese. Goslings produced in June or July, will fatten without other food than what they will have afforded them on the stubble-fields, as soon as they are ready to consume it; but, if you are in haste, give potatoes, turnips, or other roots, bruised with meal, at least, once daily. The goose is very voracious, and only requires to get plenty to eat in order to accumulate fat. Geese, fed chiefly on grass and corn, as I have described, do not, perhaps, attain the same bulk with such as are crammed; but their fat is less rank, and they are altogether much more desirable for the table.

When domesticated, highly fed, and left perfectly at ease, geese grow to a much larger size than they ever attain in a state of nature. Various arts, and often very cruel ones, have been, and are still, resorted to for the purpose of fattening them for the table, and especially for enlarging their livers, which, when thus unnaturally enlarged, and consequently diseased, are much prized by a peculiar class of epicures, although it is impossible that any part of animals which are treated in this manner can be wholesome. One mode of managing them is, to nail the webs of their feet to a board on the floor near a strong fire, to sew up the vent, and forcibly to cram them with rich food, until they are at the point of death by suffocation: by this means the liver grows to an enormous size, and the goose itself increases in weight to twenty pounds and upwards. The fat of geese principally accumulates externally; and, generally speaking, it is difficult of digestion, and therefore unwholesome. As an important department of the poultry establishment, the goose, we need hardly observe, is cultivated in almost every civilized quarter of the world, and, when under proper management, forms a profitable article of the farmer's produce, its quills, down, flesh, and even dung, being all turned to account.

Michaelmas, or stubble geese, should immediately after harvest be turned out on the wheat fields, where they pick up flesh very fast; but, when taken up to be fattened, they should be fed with ground malt mixed with water, or boiled barley and water; and, thus treated, they grow fatter than would at first be imagined, and acquire a delicate flavor. The old breeders may be plucked thrice a year, and at an interval of seven weeks, without inconvenience; but, young ones, before they are subjected to this operation, should have attained to the age of thirteen or fourteen weeks, otherwise they will pine and lose their good qualities. It is scarcely necessary to add, that the particular nature of the food, and the care that is taken of the birds, materially contribute to the value of the feathers and the down. In those neighborhoods where there is a good supply of water, they are not so subject as elsewhere to the annoyance of vermin; and they furnish feathers of a superior quality. In regard to down, there is a certain stage of maturity, which may be easily discovered, as it is then easily detached; whereas, if removed too soon, it will not keep, and is liable to be attacked by insects and their larvæ. Again, the feathers ought to be plucked, at the latest, before they are quite cold, else they will contract a bad smell, and get matted. Under proper management, and when unmolested by plucking, &c., the tame goose will live to a great age—even, it is alleged, to fourscore years, or perhaps a century.

M. Parmentier describes the French process of fattening. This consists in plucking the feathers from the belly, giving them abundance to eat and drink, cooping them up closely, and keeping them clean and quiet. The month of November is the best time to fatten geese. If the process be delayed longer, the pairing season approaches, which will prevent the birds being brought into condition.

In Poland, geese are, with this view, put into an earthen pot without a bottom, and of such a size as not to allow the bird to move; they are then fed on a paste made of ground barley, maize, buckwheat, boiled potatoes, and milk; the pot is so placed that no excrement remains in it: and the birds get very fat in about a fortnight. Even these modes appear to me cruel and unnecessary; and geese may be made fat enough for any purpose (and, indeed, too much so for the taste of most persons) by keeping them in coops in a dark place, and laying before them as much nutritious food as they can eat. This is certainly done by our continental neighbors; but then, as soon as the bird's appetite begins to flag (which is usually in about three weeks), they are forcibly crammed by means of a tin funnel, until, in about a month, the poor birds become enormously and unhealthily fat. They must then be killed, or they would die of repletion. By this process a disease of the liver is induced, in consequence of which that organ attains an unnatural size, and is regarded as a *bonne bouche* by the gourmand. Ordinary geese may readily be fattened, without cramming, to fourteen or fifteen pounds; cramming will bring up their weight to eighteen or twenty; but the excess consists of rank fat, and the flesh is deteriorated in quality, becoming actually unwholesome. The Toulouse geese readily fatten, without any cramming, up to twenty-five or even occasionally thirty pounds weight.

In some countries, the barbarous custom of plucking live geese for the sake of their feathers is resorted to. I am sorry to have to say that this cruel practice still obtains extensively. Of its barbarity, I presume I need say nothing; but I may observe, that geese so treated usually become unhealthy; many of them die; and even of such as survive, the flesh is rendered tough and unwholesome. If it be ever true, as is asserted, that the quills cast in the natural process of moulting are of inferior quality, why not clip them away close to the skin before that operation of nature begins? Then the geese will only require warmth and housing if the weather be not mild, and you will have the feathers and the geese both unimpaired in quality, and your consciences unburdened by any reminiscence of inhumanity on your part.

THE DUCK (ANAS).

THE WILD DUCK.

WILD DUCKS.—Ducks properly so called admit of a natural division into three groups, two of which have distinctive characters, while the third, which is intermediate, partakes somewhat of the character of both. This distinction is at once structural and strongly indicative of the habits of the bird, the one consisting of species which have the toes webbed together, the other of those which have the back toe loose or separate from the others. The third group alluded to, partakes more or less of the characters of each; in common language, however, the General Character Duck founded upon the Mallard or Common Wild Duck, may be considered as typical of the whole three. The older naturalists divided these into SEA DUCKS, being more perfectly web-footed, feeding principally in salt water, diving much in feeding, having a broad bill bending upwards; and POND DUCKS, haunting ponds and pools, having a straight and narrow bill, a very little hind toe, separated.

The whole tribe of Ducks, whether aquatic or more landward in their habits, find their food more by the sense of touch than by sight, and the bill is a very beautifully organized instrument for that purpose. It is covered by a sentient membrane; and the edges, which come in contact with foreign substances, are covered with papillæ, and abundantly furnished with nerves, so that when a duck dabbles in the water, the feeling in the bill enables it to distinguish eatable substances from the sludge and pebbles with which they are mixed.

The Duck in a domesticated state is an interesting and valuable bird, and an important object in rural economy. They are more intelligent than most races of ornamental poultry, and from their habit of feeding they are much less destructive, if they do not materially assist the efforts of the husbandman. When kept in a proper situation, having due access to pure water, and are fed with proper food, they are also very profitable animals; and though the flavor of their flesh is peculiar, and the fat, especially of the aquatic species, is oily and indigestible, yet they are far from unwholesome. If they have access to running streams, or even a pond of clean water, it is to be preferred, though even the ponds usually attached to farms answer very well for ducks. Where ornamental pieces of water exist in parks or pleasure-grounds attached to a gentleman's residence, ducks may be introduced with very pleasing effect, and this not only with the domesticated varieties, but even with those species which are in their natural habits the most aquatic. Even the Migratory Duck may be attracted permanently on ornamental waters and tamed. In the wild state little is known of the duck: the habits of the whole race in the breeding season are retired and silent, and as they breed in places not easily accessible to man, it may be doubted if the accounts of naturalists are to be relied upon. The males are peculiarly retiring and silent after the pairing season, and the female does not come abroad till she can launch her ducklings on the water.

THE DOMESTIC DUCK.

The duck should always find a place in the poultry-yard, provided that it can have access to water; without water it is useless endeavoring to keep these fowl; but even a very small supply will suffice. I myself have kept them with success, and fattened the ordinary duck to the weight of eight pounds, with no further supply of water than what was afforded by a large tub sunk in the ground, as I have already described when treating of poultry-yards. It must be remembered, that the flesh of these birds will be found to partake, to a great extent, of the flavor of the food on which they have been fattened; and as they are naturally very foul feeders, care should be taken for at least a week or so before killing, to confine them to select food. Boiled potatoes are very good feeding, and are still better if a little grain be mixed through them; Indian meal will be found economical and nutritive, and the best food.

Some recommend butchers' offal; but I may only warn my readers, that although ducks may be fattened on such food to an unusual weight, and thus made profitable for the market, such feeding will render their flesh rank and gross. In a garden, ducks will do good service, voraciously consuming slugs, frogs, and insects; nothing coming amiss to them; not being scratchers, they do not, like other poultry, commit such a degree of mischief in return as to counterbalance their usefulness.

The duck is very prolific; has been known to lay an egg daily for eighty-five successive days. The egg of the duck is by some people very much relished, having a rich piquancy of flavor, which gives it a decided superiority over the egg of the common fowl; and these qualities render it much in request with the pastry cook and confectioners—three duck eggs being equal in culinary value to six hen eggs. The duck does not lay during the day, but generally in the night: exceptions regulated by circumstances, will, of course, occasionally occur. While laying, the duck requires more attention than the hen, until she is accustomed to resort to a regular nest for depositing her eggs—once, however, that this is effected, she will no longer require your attendance.

The duck is a bad hatcher; she is too fond of the water, and is, consequently, too apt to suffer her eggs to get cold; she will also, no matter what sort of weather it be, bring the ducklings to the water the moment they break the shell, a practice always injurious and frequently fatal; hence the very common practice of setting duck eggs under hens. The eggs of the duck are thirty-one days in hatching; during incubation, they require no turning or other attention; and when hatched only require to be kept from water for a day or two; their first food may be boiled eggs, nettles, and a little meal; in a few days they demand no care, being perfectly able to shift for themselves; but ducks at any age are the most helpless of the inhabitants of the poultry-yard, having no weapon with which to defend themselves from vermin, or birds of prey, and their awkward waddling gait precluding their seeking safety in flight; a good stout courageous cock, and a sharp little terrier dog, are the best protectors of your poultry yard. The old duck is not so brave in the defence of her brood as the hen; but she will, nevertheless, occasionally display much spirit.

THE AYLESBURY DUCK.

There are many varieties of the Domestic Duck, the origin of which is by no means determined. White ducks have the preference with many; and of all the white ducks, the Aylesbury is the favorite. This is a large handsome bird, with plumage unspotted, and yellow legs and feet, and flesh-colored bill. Until the introduction of the variety called Rhone, or Rohan, but more probably Rouen Duck, from the town of that name on the Seine, the Aylesbury Duck was esteemed the most valuable of all; the latter bird, however, now fairly divides the honor with it, and is by some regarded as superior. The flesh of the Aylesbury duck is of a most delicate flavor, being by many compared to that of the chicken; but it is asserted that a cross between that and the Rouen Duck is superior in flavor to all others.

THE ROUEN DUCK.

The ducks of France are abundant and fine, especially in Normandy and Languedoc, where duck-liver pies are considered a great delicacy.

The Rev. Mr. Dixon seems to consider the Rouen Duck to be merely a dealer's name for the common duck. In this, however, he is scarcely justified, as it certainly possesses qualities not to be found in the common brown duck; these qualities, however, depend not on any specific differences, but on attention to a healthy mode of breeding and rearing them. The bird is very prolific, lays large eggs; and the name suits as well as another.

THE MUSCOVY OR MUSK DUCK

Does not, as some suppose, derive its name from having been brought from that country, but from the flavor of its flesh, and should more properly be termed the Musk duck, of which its other name is only a corruption; it is easily distinguished by a red membrane surrounding the eyes and covering the cheeks. These ducks, not being in esteem on account of their peculiar odor, and the unpleasant flavor of their flesh, are not worth breeding unless to cross with the common variety, in which case, let it be remarked, that the Musk drake must be put to the common duck; this will produce a very large cross, but *vice versa*, will produce a very inferior one.

MUSK DUCK.

The Musk duck is a distinct species from the common duck; and the hybrid race will, therefore, not breed again between themselves, although they are capable of doing so with either of the species from the commixture of which they sprung.

THE BLACK EAST INDIAN DUCK.

These ducks are black, and all black, feathers, legs, and bill, with a tinge of deep rich green. On a pond, mingled with the white Aylesburies, they ook extremely well, and on the spit they are more like wild duck than any other.

THE CALL DUCK.

The *bantam* of its race, usually colored like the wild mallard, but often white. This color is preferred by fowlers who use it in the decoys, as it is easily distinguished from the others. These birds have compact and elegantly rounded crests, and are very handsome.

The Aylesbury and Rouen varieties are the most valuable, and the only ones to which it is necessary to call particular attention.

The wild duck pairs strickly with a single mate; the domestic drake does not pair, and should have from four to six mates.

CHAPTER X.

THE DISEASES OF FOWL, WITH THEIR SYMPTOMS AND TREATMENT.

I MAY here premise, that when you see a fowl begnining to droop or to exhibit a deficiency of appetite, it is better at once to devote it to table use. If, however, the fowl be of great value—perhaps a Spanish cock—make an attempt to save him.

The most common diseases to which fowl are liable, are as follows:—

1. Moulting.	5. Diarrhœa.	9. Consumption.
2. Pip.	6. Indigestion.	10. Gout.
3. Inflammation.	7. Apoplexy.	11. Corns.
4. Asthma.	8. Fever	12. Costiveness.

ACCIDENTS producing fracture, bruises, ulcers, loss of feathers, &c., may, in most cases, be left to nature. When bones are broken, in most cases the patient had better be consigned to the cook. In other cases of accident the good sense of the owner will generally dictate the remedy.

MOULTING, while, as being a natural process of annual occurrence, it can scarcely be called a disease, yet must be treated of as if it really were one, from a consideration of the effects which it produces. It is most dangerous in young chickens. With adult birds, warmth and shelter are usually all that is required, united with diet of a somewhat extra stimulating and nutritious character.

Dr. Bechstein remarks, that, in a state of nature, moulting occurs to wild birds when their food is most plentiful; hence, nature herself points out that the fowl should, during that period, be furnished with an extra supply of food. After the third year the period of moulting becomes later and later, until it will sometimes happen in January or February. Of course, when this occurs, every care as to warmth should be bestowed. The use of Cayenne pepper alone, administering two or three grains made into a pill with bread, will generally suffice. Do not listen to the recommendation of ignorant or presuming quacks; if this simple treatment do not help them through, they will die in spite of all you may do.

The feathers will at times drop off fowl, when not moulting, to a very considerable extent, rendering them often nearly naked. This is a disorder similar to the mange in many other animals; and the same sort of treatment —viz., alteratives, such as sulphur and nitre, in the proportions of one quarter each, mixed with fresh butter, a change of diet, cleanliness, and fresh air in addition to this—will generally be found sufficient to effect the cure. Be careful not to confound this affection with moulting. The distinction is, that in the latter case the feathers are replaced by new ones as fast as they are cast; in the former this is not so, and the animal becomes bald. Mr. Martin relates an anecdote which would indicate that fear has influences as great upon birds as on the human being. "A cock," he says, " belonging to a friend, was dreadfully frightened by a dog, and became white, but recovered his natural plumage at the next moult. A black Poland cock, being seized near the house by a fox, his screams being heard, he was rescued, desperately wounded, with the loss of half his feathers. In time the remainder of his feathers came off, and he became perfectly white."

Pip.—A disease to which young fowl are peculiarly liable, and that, too, chiefly in hot weather.

The symptoms are—a thickening of the membrane of the tongue, especially towards its tip. This speedily becomes an obstruction of sufficient magnitude to impede the breathing; this produces gasping for breath; and at this stage the beak will often be held open. The plumage becomes ruffled and neglected, especially about the head and neck. The appetite gradually goes; and the poor bird shows its distress by pining, moping, and seeking solitude and darkness.

The cause of this disease is want of clean water and feeding too much upon hot exciting food. Dr. Bechstein considers it to be analogous to the influenza of human beings. Theories respecting its nature are numerous and of very little practical importance.

Cure.—Most writers recommend the immediate removal of the thickened membrane. I do not like this. Mr. Martin in his excellent work, recommends that the tongue be cleansed by applying a little borax dissolved in tincture of myrrh, by means of a camel-hair pencil, two or three times a-day. We would rather anoint the part with fresh butter or cream. Prick the scab with a needle, if you like; and give internally a pill about the size of a marble, composed of:—Garlic, and Horse-radish scraped, in equal parts, as much Cayenne pepper as will outweigh a grain of wheat. Mix this with fresh butter, and give it every morning—keeping the fowl warm. Keep the bird supplied with plenty of fresh water; preserve it from molestation, by keeping it by itself, and you will generally find it get well if you have taken the disease in time. Do not let any one, equally ignorant and cruel, persuade you to cram the mouth with snuff after having torn off the thickened membrane with your nail. This is equally repugnant to humanity and common reason. Forcing tobacco-smoke down the bird's throat is advised; and when, as sometimes is the case, the disease depends on the presence of a worm, then it is most successful.

INFLAMMATION.—Most of the diseases to which poultry are subject may be traced to inflammation exhibiting itself in some part of the system.

INFLAMMATION OF THE TRACHEA.—The disease to which this term is improperly applied is an inflammation of the tail-gland. The true roup is a disease extremely analogous to influenza in man, or even more so to the well-known distemper among dogs; and, in some forms, perhaps, to the glanders of the horse, and is sometimes termed Gapes and sometimes Roup or Croup.

The symptoms are—difficulty of breathing, constant gaping, dimness of sight, lividity of the eyelids, and the total loss of sight; a discharge from the nostrils, that gradually becomes purulent and fetid: appetite has fled; but thirst remains to the most aggravated extent. Sometimes this disease appears to occur independently of any obvious cause; but dirt, too hot feeding and want of exercise are amongst the most usual.

The remedies recommended are various. Mr. Martin prescribes one grain calomel made up with bread into a pill, or, if preferred, two or three grains Plummer's pills (*pil. hydr. Submur. co.* Lond. Pharm.), after which let flour of sulphur be administered mixed with a little ginger, mixed barley meal reduced to a paste, and the mouth well washed in a weak solution of chloride of lime. In the mean time, let the bird be kept in a dry, warm, well ventilated apartment, and apart from the other fowl. When the bird dies of this disease, the trachea will be found replete with narrow worms about half an inch in length imbedded in slimy mucus. This singular worm is the *distoma lineare*, a long and short body united, the long body being the female, the short the male; they are permanently united, otherwise they are quite perfect in themselves. Mr. Martin is uncertain if these worms are the cause or consequence of the disease; but it is certain when they have once established themselves, their removal is necessary to give the bird a chance of recovery. This is sometimes done by means of a feather, neatly trimmed, which is introduced into the windpipe, and turned round once or twice, and then drawn out; this will dislodge some of the worms if dexterously performed, and with some knowledge of the anatomy of the parts.

Spirit of turpentine in rice, and afterwards a little salt in the water, have been given successfully.

My treatment would merely be warmth and cleanliness, as matters of course; and for pellets—

Powdered gentian,	. .	1 part,
Powdered ginger,	. . .	1 "
Epsom salts,	. .	1½ "
Flour of sulphur,	. . .	½ "

Make up with butter, and give every morning.

If the discharge should become fetid, the mouth, nostrils, and eyes may be bathed with a weak solution, composed of equal parts of chloride of lime and acetate of lead. Fomentation with an infusion of chamomile flowers is also highly beneficial.

The other affection—that improperly passed under this name—viz., swelling of the tail-gland—may be treated as a boil. If it become inconveniently hard and ripe, let the pus or matter out with a penknife, and it will soon get well.

INFLAMMATION OF THE LUNGS is attended by quick breathing with a rattle, an audible dullness, disorder of plumage, vacancy in the eye, and general indisposition. Bleeding, the natural remedy for such symptoms, is out of the question, for how is a bird to be bled, and where?

INFLAMMATION OF THE HEART.—A fatal disease among poultry, and only detected by examination after death. The patient appears to droop, refuses to eat, retires to roost, and is found dead in the morning. In this case, the peritoneal membrane exhibits indications of active inflammation.

INFLAMMATION OF THE MUCOUS MEMBRANE.—Generally proceeds from aggravated diarrhœa. The bird is severely purged, and the evacuations become more or less tinged with blood, and death ensues unless a speedy remedy is applied. Damp and improper food are the cause of the disease. The remedy, to be successful, must be administered early; first, give a small quantity of castor oil; this will clear the bowels of irritating secretions; afterwards, give doses of *hydrargyrum cum creta*, (Lond. Pharmacopœia), with rhubarb and laudanum, as follows:—

Hydr. cum creta	. .	3 grains.
Rhubarb,	. . .	2 or 3 grains.
Laudanum,	. .	2, 3, or 4 drops.

Mix in a teaspoonful of gruel, and give twice a day.

ASTHMA is characterised by gaping, panting, and difficulty of breathing.

We need not go far to seek for a cause. Our poultry are originally natives of tropical climates; and, however well they may appear climatized, they, nevertheless, require an equable temperature, unaided by artificial means. Hence coughs, colds, catarrh, asthma, pulmonary consumption, arise from variable climate.

CURE.—Warmth, with small repeated doses of hippo powder and sulphur mixed with butter. The addition of Cayenne pepper will be an improvement.

DIARRHŒA is occasioned by damp, and sometimes by improper food. Remove the bird into dry quarters; change the food; if it become very severe, give chalk; add a little starch, mixed with Cayenne, to porridge, and give it moderately warm.

INDIGESTION.—Caused by over-feeding and want of exercise. Lessen the quantity of food; turn the fowl into an open walk, and give some powdered gentian and Cayenne in the food.

APOPLEXY.—Symptoms—Staggering, shaking of the head, and a sort of tipsy aspect. Some persons have, from ignorance of the true cause of this affection, treated it as proceeding from intestinal irritation, and prescribed castor-oil with syrup of ginger, &c. Scanty food, and that of light quality, and the application of leeches to the back of the neck, constitute, in my opinion, the only effectual remedy. Perhaps, however, it is better to have the poor bird at once handed over to the cook.

PARASITES IN FOWLS.—The insects which infest animals of all kinds, more especially domesticated ones, are the bane of their existence. In poultry they are particularly obnoxious, and the utmost possible cleanliness and frequent lime-washing and fumigation, are necessary to keep them in proper condition. White precipitate powder, applied with a small camel-hair pencil, in small quantities, will destroy lice and other parasites.

Like the domestic fowl the peacock has also its parasites in the *Goniodes pulcicornis*. After the death of the bird the insect may be found congregated in numbers about the base of the beak and crown of the head. Mr. Denny was induced to examine all the genera of domesticated birds, and he found on the Turkey *Lipeurus polytrapezius* as a common parasite; the *Goniodes stylifer* is also frequent in the head, neck, and breast. Over the domestic fowl he found three species of parasitic *Gonoides dissimilis* of rare occurrence. *Lipeurus variabilis* preferring the primary and secondary feathers of the wing, among the ribs of which they move with great celerity. *Menopon pallidum* he also found in great abundance on the domestic fowl; and, as a general rule, he observed that when two or more species frequented the same species of bird, each had its own locality.

The remedy in all cases is cleanliness, and when the fowls are over infested, fumigation and a plentiful sand bath of clean, dry and rough sand; for the white precipitate powder, named above, is poisonous, and only fit to be used on very young birds, which have not yet learned the art of preening their feathers with their bill.

FEVER.—Fowl are frequently subject to febrile affections. The mode of treatment is simple—light food, and little of it; change of air; and, if necessary aperient medicines—such as castor-oil, with a little burnt butter.

CONSUMPTION I regard as incurable; but, if anything will do good, it is change of air and warmth.

Gout.—Its effects are obvious. Pellets of Colchicum may be used; but if you had, as you should have done, killed your fowl before they became so old, it would have been more rational. They are now past use. Sulphur may also be found useful.

Corns.—These may generally be extracted with the point of a penknife. If ulcerated, as will often occur when neglected, touch with lunar caustic, and you may thus succeed in establishing healthy granulations.

Costiveness.—This affection will, in general, yield to castor-oil and burnt butter. The diet should be sparing. Thin porridge will be found useful.

In the case of fractures, put the fowl to death without loss of time. The same may be said of bruises. By this you not merely avoid some loss, but save the poor bird much suffering.

The accidental stripping of the feathers must not be confounded with the mangy affection already treated of. The difference will be seen by examining the state of the skin where it is exposed.

Ulcers may be kept clean, dressed with a little lard, or washed with a weak solution of sugar of lead, as their aspect indicates. If sluggish, touch with bluestone.

CHAPTER XI.

CAPONIZING.

The objects proposed in converting a cock into a capon are the following:—his natural fierceness is quelled; he becomes placid and peaceful; his pugnacity has deserted him; he no longer seeks the company of the hens; he grows to a far larger size than he otherwise would have done; he acquires flesh with much greater rapidity, and that flesh is peculiarly white, firm, and succulent, and even the fat is perfectly destitute of rankness. To these advantages another may, perhaps, be added—viz., the capon may, by a little management, be converted into an admirable nurse, and will be found particularly valuable, in this respect, to parties using the *eccalobeion*, or hatching-machine.

The process has been made a subject of much unnecessary mystery, and, I regret to add, of much unnecessary cruelty. In point of fact, caponizing is an extremely simple affair, which the country henwives in France perform with facility and certainty. The practice of the French country women is to select the close of the spring, or the beginning of autumn, as well as fine weather, for the performance of their work. The parts necessary to be removed being fixed in the abdomen, and attached to the spine at the region of the loins, it is absolutely necessary to open the abdominal cavity for the purpose of their extraction. The bird should be healthy, fasting, and about three months old. He is then to be secured by an assistant, upon his back, his belly upwards, and his head down, that the intestines, &c., may fall up towards the breast; the tail is to be towards the operator. The right leg is then carried along the body, and the left brought backwards, and held in this position, so as to leave the left flank perfectly bare, for it is there that the incision is to be made. The said incision is to be directed from before backwards, transversely to the length of the body, at the middle of the flank and slightly to the side between the ends of the breast-bone and the vent. Having plucked away the feathers from the space where it is intended to make the incision, you take a bistoury or a scalpel, and cut through the skin, abdominal muscles, and peritoneum; it is better to do this at two or more cuts, in order to avoid the possibility of wounding the intestines—a casualty that would, in most cases, be attended with fatal results. The intestines present themselves at the orifice; but you must not suffer them to come out; on the contrary, you press them gently aside, so as to have room for action. I may observe, that the incision should have been sufficiently large to admit of the forefinger, previously well oiled, being passed into the abdomen, and carried carefully towards the lumbar region of the spine: you will there find what you are in search of. You first reach the left substance, which you detach with your nail, or with your finger bent hook-fashion; then you arrive at the right, which you treat similarly—bring both substances forth; you finally return the intestines, sow up the wound with a silk thread—a very few stitches will suffice—and smear the place with a little fresh butter. The comb of the capon does not grow to any size, and always retains a pallid color.

The process having been performed as above described, the bird is placed in a warm house, where there are no perches, as, if such appliances were present, the newly-made capon might very probably injure himself in his attempts to perch, and perhaps even tear open the sutures, and possibly occasion the operation, usually simple and free from danger, to terminate fatally. For about a week, the food of the bird should be soft oatmeal porridge, and that in small quantities, alternated with bread steeped in milk; he may be given as much pure water as he will drink; but I recommend that it be tepid, or at least that the chill be taken off it. At the end of a week, or, at the farthest, ten days, the bird, if he has been previously of a sound vigorous constitution, will be all right, and may be turned out into the walk common to all your fowl.

The Malays are particularly adapted for caponizing, and, when properly fattened, at a suitable time after the operation, attain a bulk and weight that would surprise such persons as have never seen a caponized specimen of that breed, the birds, in fact, rivalling the finest turkeys.

94 DOMESTIC FOWL.

An operation of a similar nature is performed upon hens, either before they have begun to lay, or after they have ceased to do so, for the purpose of preventing them from laying in future. This renders them, as the other does the cock, more susceptible of taking flesh, and that of a finer quality than ordinary. It is proper to remind the reader that, of course, when it is deemed advisable thus to deprive a hen of the power of reproduction, such a one should always be selected as presents deformities or other defects that ought to render her unfit for breeding purposes.

The caponizing of pullets is performed in much the same manner as in the case of cocks. The oviduct is found towards the loins, and is extracted in the same manner as already described in the former case. Some French writers, however, and Schreger amongst the first, state, that in the case of pullets or hens the operation is unnecessary, it being only required to make a small incision just above the vent, on a little eminence that will be perceived in that place; then, by repeated pressure, you cause the protrusion of the uterus—a little whitish body; this is cut away, the wound heals of itself, and nothing further is required.

When necessary, in either case, to employ sutures for the purpose of closing the wound, great care must be taken to avoid involving the intestines in the stitches. I warn the operator that, if he be tedious in the performance of his work, the chances are greatly against his success. Whoever proposes to caponize should acquire dexterity of manipulation by practising on the dead bird, before he endeavors to use his knife upon the living: when such

INSTRUMENTS FOR CAPONISING.

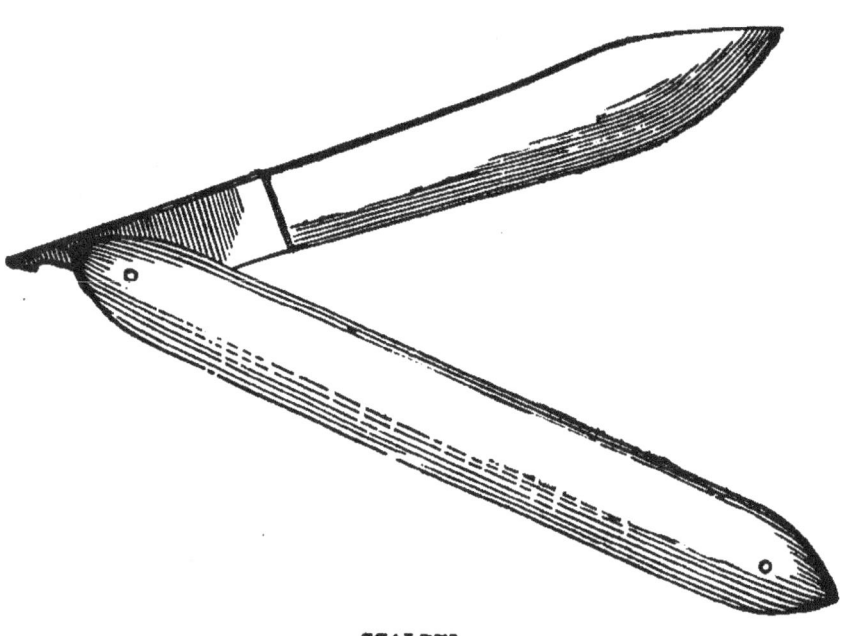

SCALPEL.

CAPONISING.

RETRACTOR.

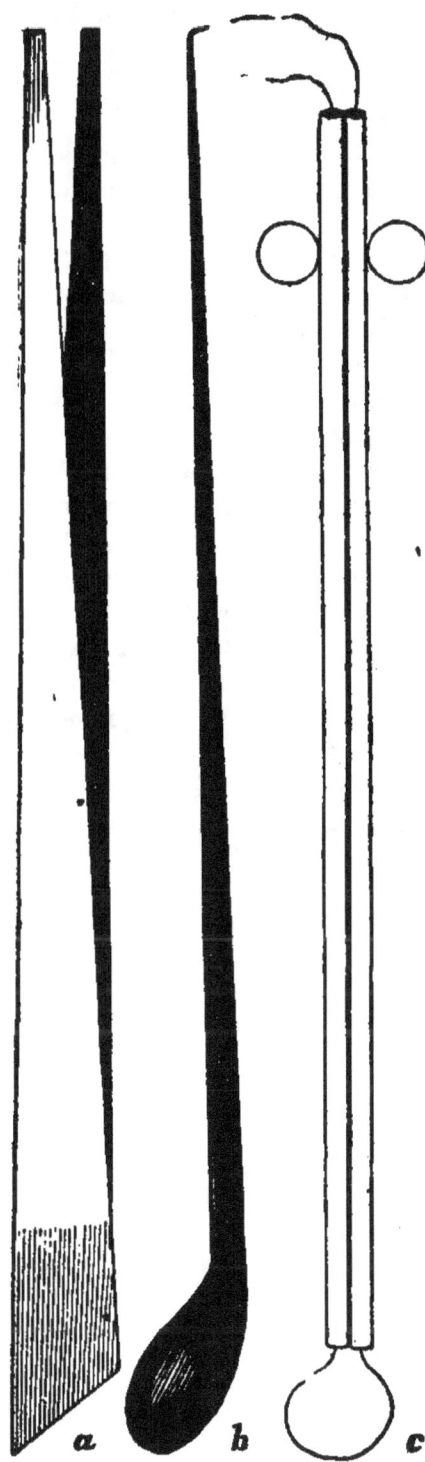

FORCEPS, SPOON AND CANULA.

INSTRUMENTS FOR CAPONISING.

These are, a *scalpel*, for cutting open the fowl, and cutting the thin skin which envelops the testicles; a silver *retractor*, for stretching open the wound wide enough to operate within; a pair of *spring forceps*, denoted by the letter *a*, having a sharp cutting edge, like that of a chisel, with a bevel half an inch in its greatest width, for making the incision and securing the thin membrane covering the testicles; a *spoon-shaped tool*, marked *b*, with a sharp hook at one end for pushing and removing the testicles, adjusting the loop, and to assist in tearing open the tender covering of the testicles; and a *double silver canula*, marked *c*, for containing the two ends of horse hair or fibre constituting the loop, to be passed around the testicle to draw it out.

precautions are used, the operation will be divested of much of its apparent cruelty; and if to be resorted to at all, every precaution should be taken to outrage humanity as little as possible.

Sometimes, but rarely, this operation is performed on turkeys, geese, and ducks; the reason why it is performed so rarely on these birds is, that, from the great plumpness of their bodies, what we want to arrive at is farther from our reach, the operation, of course, so much the more difficult, and the probability of success so much the more remote.

The capon is so very much disposed of itself to take on flesh that it will, in general, attain to sufficient condition in the yard, or about the barn door. Sometimes, however, it is deemed advisable to cram him. This practice induces rapid growth in little time, a very delicate quality of flesh (I except the caponized gander from this), and also causes him to fetch a higher price in the market. When it is considered desirable to cram a capon, he is taken and placed in a dark and quiet house, or coop, so small that he shall be unable to exercise; he may then be fed with pellets of meal and milk. Pea-meal or bean-meal will be found to impart a fine flavor to the flesh, but if this description of food be found too binding, let pellets of barley-meal be given, till the undesired effect is removed; the bird should be left as much food as he will eat, and should, besides, be crammed at least three times a day. In three weeks he will be ready for use. It may not be amiss here to remind the reader that the droppings of the bird are almost, if not quite, as valuable as guano for the purpose of manure.

A little dish of fine gravel or coarse sand, left in the feeding trough, will be relished by the birds, will promote digestion, and will, of course, thus aid in conducing to their rapid fattening.

PUBLISHER'S ADVERTISEMENT.

The Publisher, having found the want of small, cheap Books, of acknowledged merit, on the great topic of farming economy, and meeting for those of such a class a constant demand, offers, in his Rural Handbooks, of which this is one, works calculated to fill the void.

He trusts that a discerning Public will both buy and read these little Treatises, so admirably adapted to all classes, and fitted by their size for the pocket, and thus readable at the fireside, on the road, and in short everywhere.

C. M. SAXTON,
Agricultural Book Publisher.

EDITOR'S PREFACE.

THE American Publisher placed this little book in the hands of the editor to be Americanized. Mr. Milburn, its author, is a resident of Yorkshire, England, a county whose cattle are almost wholly of the short horn kind. While he does ample justice to the favorite breed of his own region, he much underrates the Devons and Herefords. It is unfortunately the case that in England the breeders of particular races of animals admit little or no merit in other varieties. Due allowance must be made for the author's position in a short horn cattle region, and his consequent partialities.

Cattle, milk, butter, and cheese-making are the same essentially in all countries. Variations of climate and soil make different breeds of beasts preferable in different regions, and the same causes act to somewhat change the processes of producing butter and cheese. Yet a good treatise on these subjects will suit all countries and all varieties of breeds.

This little book contains much valuable matter, in a compact form and at a low price, that is nowhere else so accessible and so reasonable. The American editor has adapted it to the American farmer and breeder, preserving the views and opinions of the author, correcting obvious errors in fact, and rendering this edition of more value to the American reader than the English one.

In America it has been found, that the Devons, while suited to all latitudes, are better fitted for the climate of the extreme south and the extreme north than any other breed of cattle. In Georgia and Canada, they are superior to all others either for milk or feeding. In a few years all New England will be occupied with this breed, as the one best adapted to its pastures and its climate.

The short horn and the Alderney are by experience shown to be adapted only to the more temperate portions of our country. Within this range the short horn has no equal as a whole, for beef and milk.

The American breeder and dairyman should therefore choose his breed in reference to his locality. In doing this he should confine himself to the short horn and Devon races, when he designs to breed with a view to both beef and milk. In merely grazing regions, where the dairy forms no object, he may select also the Herefords as a good beef and feeding race, and adapted to all regions whose pasture and climate suit the short horn. So far as experiments have been tried, the Ayrshire breed may be said, in general, to have failed, and should therefore be avoided. The Alderney is only to be commended to the amateur, or those keeping a single cow; she gives a small quantity of very rich milk, that furnishes a great luxury in its cream and butter; but while these are rich, they are too small in quantity, and the milk is very defective in its cheese-making property.

The American breeder will thus see the propriety of being guided by the experience of those who have tested in their own climate and on their own soils, the different breeds, and shown those which are best fitted for both.

AMBROSE STEVENS.

www.ingramcontent.com/pod-product-compliance
Lightning Source LLC
Chambersburg PA
CBHW081118240526
45470CB00019B/2611